T0336101

Laboratory Manual for Groundwater, Wells, and Pumps

The over-exploitation of groundwater and marked changes in climate over recent decades have led to unacceptable declines in groundwater resources. Under the likely scarcity of available water resources in the near future, it is critical to quantify and manage the available water resources. With increasing demand for potable water for human consumption, agriculture, and industrial uses, the need to evaluate the groundwater development, management, and productivity of aquifers also increases. *Laboratory Manual for Groundwater, Wells, and Pumps* serves as a valuable resource and provides a multi-disciplinary overview for academics, administrators, scientists, policymakers, and professionals involved in managing sustainable groundwater development programs. It includes practical guidance on the measurement of groundwater flow, soil properties, aquifer properties, wells and their design, as well as the latest state-of-the-art information on pumps and their testing, and groundwater modeling.

Features:

- Covers basics of groundwater engineering, advanced methodologies, and their applications and groundwater modeling
- Examines groundwater exploration, planning and designing, and methods for formulating strategies for sustainable management and development
- Serves as a reference for practitioners on practical applications and frequently occurring issues of groundwater investigations, development, and management.

Laboratory Manual for Groundwater, Wells, and Pumps

Rohitashw Kumar, Vijay P. Singh, and
Munjid Maryam

CRC Press
Taylor & Francis Group
Boca Raton London New York

CRC Press is an imprint of the
Taylor & Francis Group, an **informa** business

Designed cover image: Shutterstock

First edition published 2023
by CRC Press
6000 Broken Sound Parkway NW, Suite 300, Boca Raton, FL 33487-2742

and by CRC Press
4 Park Square, Milton Park, Abingdon, Oxon, OX14 4RN

CRC Press is an imprint of Taylor & Francis Group, LLC

© 2023 Rohitashw Kumar, Vijay P. Singh, and Munjid Maryam

ISBN: 978-1-032-33433-2 (hbk)
ISBN: 978-1-032-33467-7 (pbk)
ISBN: 978-1-003-31975-7 (ebk)

DOI: 10.1201/9781003319757

Typeset in Times
by codeMantra

Contents

PART 1   *Measurement of Groundwater Flow*

PART 2   *Measurement of Soil Properties*

PART 3 Measurement of Aquifer Properties

PART 4   Wells and Their Design

PART 5   Pumps and Their Testing

PART 6   *Groundwater Modeling*

Preface

Agricultural productivity growth is vital for economic and food security outcomes which are threatened by climate change. In the present times, changing climate is a burning issue. Climate change may be limited to a specific region or may occur across the earth. One of the nation's most significant natural resources is ground water. Approximately 40% of the water needed for all uses, excluding hydropower generation and cooling electric power plants, comes from this source. Remarkably, the existence of groundwater is not only poorly known, but it is also the topic of many common myths for a resource that is so frequently used and crucial to the health and prosperity of a nation. One common misunderstanding is that groundwater is found in underground rivers that resemble surface streams and that certain people can sense their presence. These myths and others have made it difficult to develop and conserve groundwater and have hurt efforts to preserve its quality.

Everyone, from the rural homeowner to managers of industrial and municipal water supplies to heads of Federal and State water regulatory agencies, must become more knowledgeable about the occurrence, development, and protection of groundwater in order for the nation to reap the greatest benefits from its groundwater resource. These organizations' interests, as well as the needs of hydrologists, well drillers, and other people involved in the research and development of groundwater supplies, have all been taken into consideration while creating this manual.

The present practical manual will be carried out for the development and management of groundwater resources for providing the constant scientific output using advanced techniques and modeling, emphasizing the sustainable groundwater development and management programs. Over-exploitation of groundwater and marked changes in climate over the last couple of decades have imposed immense pressure on the global groundwater resources. Under the foreseen scarcity situation of available water resources in the near future and its intimidating threats, it has become very crucial for the earth scientists as well as planners to quantify and manage the available water resources. As demand of potable water increased manifold across the globe for human consumption, agriculture, and industrial uses, the need to evaluate the groundwater development, management, and productivity of aquifers has also increased. The proposed manual covers recent geographic information system-based studies, advanced methodologies with their applications, and groundwater modeling. The book stands significant in highlighting prominent outputs in groundwater exploration planning and designing, and formulating further strategies under the sustainable management and development theme.

This book is a collection of chapters providing a multi-disciplinary overview for academics, administrators, scientists, policymakers, social science, professionals, and Non-Governmental Organizations (NGOs) involved in the lined departments running sustainable groundwater development and management programs. The content of the book covers themes like groundwater occurrence and movement, measurement of groundwater flow, measurement of soil properties, measurement of aquifer properties, wells and their design, pumps and their testing well-aquifer interactions, groundwater investigations, aquifer test analyses, predicting aquifer yield, data collecting, and geophysical investigations. Additionally, discussions have included permeability tests, well design, well drilling, pumps, and groundwater modeling, among other topics. There is also a substantial bibliography available.

The main purpose of this manual is to formulate the plan for the development and maintenance of groundwater from India and other countries having a similar setup of natural resources, particularly water. The proposed book, *Laboratory Manual on Groundwater Wells and Pumps*, attempts to fill this need.

There are six sections on the fundamentals of groundwater hydrology, which are organized in descending sequence from the most fundamental to a review of the techniques used to estimate aquifer yield to a discussion of common issues with the operation of groundwater supplies. Each segment is composed of a concise text and one or more illustrations, such as maps or drawings that highlight the key ideas discussed in the book.

Section 1 deals with the measurement of groundwater and is subdivided into: Verification of Darcy's law which determines how groundwater moves. It describes a fluid's capacity to permeate a porous medium like rock, estimating groundwater balance and study of artificial groundwater recharge structures which specifies that whatever water enters a designated area ought to go into storage within its bounds, be consumed there, be exported there, or flow out on the surface or underground within a specified time period. It also deals with water table contour maps. It describes how water table estimations that are taken at any corresponding time of the year are used to draw a water table contour map. The water table contour map helps in portraying the groundwater flow direction.

Section 2 deals with the measurement of soil properties and is subdivided into: Sieve analysis for gravel and well screens design which helps to understand the properties of the water-bearing formation revealed by sieve examination of samples acquired during the drilling of test holes or production wells. The findings serve as a foundation for making judgments about well screen specification and gravel pack design.

Section 3 deals with the measurement of aquifer properties and is subdivided into: Estimation of specific yield and specific retention, Evaluation of hydraulic properties of aquifer by Theis method, Evaluation of hydraulic properties of aquifer by the Cooper-Jacob Method, Evaluation of hydraulic properties of aquifer under unsteady state condition by the Chow method, and Evaluation of hydraulic properties of aquifer by recovery test and Leaky and non-leaky aquifers.

Section 4 deals with wells and their design and is subdivided into: Testing of well screen, Study of different drilling equipment and drilling of a tube well, Measurement of water level and drawdown in pumped wells, Well design under confined and unconfined conditions, Study of well losses and well efficiency, and Computation of interference of wells.

Section 5 deals with pumps and their testing and is subdivided into: Study of radial flow, mixed flow, multistage centrifugal pumps, turbine, propeller and other pumps, Study of installation of centrifugal pump, testing of centrifugal pump, and study of cavitations, Study of performance characteristics of hydraulic ram, Study of deep well turbine and submersible pumps, and Analysis of pumping test data.

Section 6 deals with groundwater modeling and is subdivided into: Groundwater computer simulation models. It helps us understand how groundwater models use numerical circumstances to represent the groundwater flow and transport processes. The direction of flow, the aquifer's geology, the heterogeneity or anisotropy of the sediments or bedrock within the aquifer, the foreign substance transport mechanisms, and the substance reactions are typically included in these hypotheses.

An appendix is provided at the end which is the description of conversion factors of units from one system to another. Glossary is also provided to explain the various terminologies associated with groundwater, wells, and pumps.

Where the phrases are initially presented, definitions of groundwater terminologies are provided. A section on numerical is also provided at the conclusion of every chapter for individuals who need to revisit some of the mathematical equations used in groundwater hydrology.

The precise purpose, expectations, and learning objectives of the groundwater, wells, and pumps are defined and distinguished in this manual. This lab manual is intended to go along with a college course that introduces students to groundwater flow, hydrology, aquifer and their properties, wells and their design, and study and installation of pumps and groundwater simulation models. This manual will also serve as a reference for field staff on the more practical elements and frequently occurring issues of groundwater investigations, development, and management. This manual will be helpful in boosting groundwater hydrology productivity, profitability, and appropriateness.

As authors, we realize that we have just begun to scratch the surface with some of the recent advances in groundwater, wells, and pumps. This manual discusses examples of groundwater hydrology and is based on experiences at different regions throughout the world. It is hoped that this book will be helpful in boosting groundwater hydrology productivity, profitability, and appropriateness.

Rohitashw Kumar

Vijay P. Singh

Munjid Maryam

Acknowledgments

I sincerely acknowledge the efforts of Professor Vijay Pal Singh (co-author), distinguished Professor, Department of Biological and Agricultural Engineering, Texas A&M University, College Station, Texas, USA, for his support and guidance to bring this manuscript in final refined shape, constructive criticism, and utmost cooperation at every stage during this work. He is always an inspiration for us.

I am grateful to Er. Munjid Maryam (co-author), who put great efforts and worked day and night to bring this manuscript in a refined shape.

I am grateful to my students Dr. Sakeel Ahmad Bhat, Er. Zeenat Farooq, Er. Tanzeel Khan, Er. Dinesh Vishkarma, Er. Faizan Masoodi, Er Khilat Shabir, Er, Noureen, and Er. Mahrukh for providing all necessary help to write this book.

I am highly thankful to Dr. Ishfaq for his valuable contributions in writing of this book.

I am highly obliged to Hon'ble Vice Chancellor Prof. N. A. Ganai for his support and affection and encouragement always for innovation. I wish to extend my sincere thanks to College of Agricultural Engineering and Technology, SKUAST-Kashmir, for their support and encouragement.

I express my regards and reverence to my late parents as their contribution in whatever I have achieved till date is beyond expression. It was their love, affection, and blessed care that have helped me to move ahead in my difficult times and complete my work successfully. I thank my family members, who have been a source of inspiration always. I thankfully acknowledge the contribution of all my teachers since schooldays, for showing me the right path at different steps of life.

I consider it a proud privilege to express my heartfelt gratitude to ICAR – All India Coordinated Research Project on Plastic Engineering in Agriculture Structures & Environment Management for providing all facility to carry out this project at SKUAST-Kashmir, Srinagar.

I sincerely acknowledge with love, the patience, and support of my wonderful wife Reshma. She has loved and cared for me without ever asking anything in return, and I am thankful to God for blessing me with her. She has spent the best and the worst of times with me, but her faith in my decisions and my abilities has never wavered. I would also like to thank my beloved daughter Meenu and son Vineet for making my home lovely with their sweet activities.

Finally, I bow my head before the almighty God, whose divine grace gave me the required courage, strength, and perseverance to overcome various obstacles that stood in my way.

Rohitashw Kumar
Associate Dean,
College of Agricultural Engineering and Technology,
SKUAST- Kashmir, Srinagar

Authors

Dr. Rohitashw Kumar (B.E., M.E., Ph.D.) is Associate Dean and Professor in College of Agricultural Engineering and Technology, Sher-e-Kashmir University of Agricultural Sciences and Technology of Kashmir, Srinagar, India. He worked as Professor Water Chair (Sheikkul Alam ShiekhNuruddin Water Chair), Ministry of Jal Shakti, Government of India, at the National Institute of Technology, Srinagar (J&K) for three years. He is also Professor and Head, Division of Irrigation and Drainage Engineering. He obtained his Ph.D. degree in Water Resources Engineering from NIT, Hamirpur, and Master of Engineering Degree in Irrigation Water Management Engineering from MPUAT, Udaipur. He got Leadership Award in 2020, Special Research Award in 2017, and Student Incentive Award in 2015 (Ph.D. Research) from the Soil Conservation Society of India, New Delhi. He also got the first prize in India for the best M.Tech. thesis in Agricultural Engineering in 2001. He has published over 129 papers in peer-reviewed journals, more than 25 popular articles, four books, four practical manuals, and 25 book chapters. He has guided two Ph.D. students and 17 M.Tech. in soil and water engineering. He has handled more than 14 research projects as a principal or co-principal investigator. Since 2011, he has been Principal Investigator of ICAR – All India Coordinated Research Project on Plastic Engineering in Agriculture Structural and Environment Management.

Prof. Vijay P. Singh is a Distinguished Professor, a Regents Professor, and the inaugural holder of the Caroline and William N. Lehrer Distinguished Chair in Water Engineering at the Texas A&M University. His research interests include surface-water hydrology, groundwater hydrology, hydraulics, irrigation engineering, environmental quality, water resources, water–food–energy nexus, climate change impacts, entropy theory, copula theory, and mathematical modeling. He graduated with a B.Sc. in Engineering and Technology with emphasis on Soil and Water Conservation Engineering in 1967 from the U.P. Agricultural University, India. He earned an M.S. in Engineering with specialization in Hydrology in 1970 from the University of Guelph, Canada; a Ph.D. in Civil Engineering with specialization in Hydrology and Water Resources in 1974 from the Colorado State University, Fort Collins, USA; and a D.Sc. in Environmental and Water Resources Engineering in 1998 from the University of the Witwatersrand, Johannesburg, South Africa. He has published extensively on a wide range of topics. His publications include more than 1,365 journal articles, 32 books, 80 edited books, 305 book chapters, and 315 conference proceedings papers. For his seminar contributions, he has received more than 100 national and international awards, including three honorary doctorates. Currently, he serves as Past President of the American Academy of Water Resources Engineers and the American Society of Civil Engineers (ASCE), and previously he served as President of the American Institute of Hydrology and Cahir, Watershed Council, ASCE. He is Editor-in-Chief of two book series and three journals, and serves on the editorial boards of more than 25 journals. He has served as Editor-in-Chief of three other journals. He is a Distinguished Member of the American Society of Civil Engineers, an Honorary Member of the American Water Resources Association, an Honorary Member of International Water Resource Association, and a Distinguished Fellow of the Association of Global Groundwater Scientists. He is a fellow of five professional societies. He is also a fellow or member of 11 national or international engineering or science academies.

Er. Munjid Maryam did B.Tech. in Agricultural Engineering and M.Tech. in Soil and Water Engineering from College of Agricultural Engineering and Technology, SKUAST-Kashmir. Currently, she is pursuing Ph.D. in Soil and Water Engineering in sandwich mode from College of Agricultural Engineering and Technology, SKUAST-Kashmir and Western Sydney University, Australia. She has published various research and review articles in the area of water resources engineering. She is the recipient of the Best Speaker Award at the National Environment Youth Parliament 2022 held in New Delhi, India. Furthermore, she is also a recipient of EIR fellowship issued by BITS Pilani, India.

Part 1

Measurement of Groundwater Flow

1 Verification of Darcy's Law

Objective: Verification of Darcy's Law

1.1 THEORY

Darcy's law determines how groundwater moves. It describes a fluid's capacity to permeate a porous medium like rock. It is based on the notion that the rate of flow or flux in between two points in the saturated medium is proportionate to the variation in pressure between them, their distance, and the interconnectedness of flow routes as well as fluid properties, such as density and viscosity. Permeability is used for measuring the interconnectivity. The constant of proportionality equals the product of two factors: the function of fluid density and viscosity, and the function of permeability which is directly proportional to the square of the grain diameter.

During geologic evolution, rock is deposited in layers in the subsurface. The permeability of rocks controls fluid flow inside and between the strata and must be determined together in vertical and horizontal directions. Shale, for instance, has substantially lower vertical permeabilities than horizontal permeabilities (assuming flat-lying shale beds). This makes the fluid passage up and down through a shale bed challenging, whereas fluid flow from side to side is significantly easier. There are natural flow paths in horizontal bedding planes in shales, which make it easier for water to flow through in comparison to vertical flow where there are few flow pathways.

Flow between zones is unlikely to occur when the difference in pressure between a hydraulically cleaved zone and a freshwater aquifer is not elevated, the space between the zones is relatively extensive, and the presence of rocks with restricted vertical permeabilities between cavernous and shallower expanse. An anomaly to this concept is a separate flow pathway, like an open borehole or a succession of faults or joints which bisect both the cleaved zone and the freshwater aquifer. The pressure differential and distance are the decisive elements in migration of fluid from the lower zone to the upper zone in each of these situations.

Darcy's law can be written as:

$$Q = -KA\frac{dh}{dl} \tag{1.1}$$

where

Q: Rate of water flow (volume per time);
K: Hydraulic conductivity;

DOI: 10.1201/9781003319757-2

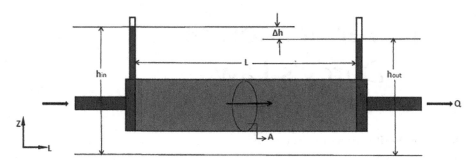

FIGURE 1.1   Schematic representation of Darcy's law.

A: Column cross-sectional area; and
dh/dl: Hydraulic gradient, that is, the change in head over the length of interest.

Figure 1.1 is a diagrammatic expression of Darcy's law.

Darcy's law is applied when calculating flow of fluid from a hydraulically cleaved zone to a freshwater zone. For ascertaining how hydraulic fracture fluids reach a freshwater zone and determining the states which govern the movement of fluid from one zone to another zone, this law can be taken into account.

The unit darcy is dependent upon a lot of different unit systems. A medium that has a permeability of 1 darcy permits a fluid with a viscosity of 1 cP (1 mPa·s) to flow at 1 cm³/s across a 1 cm² area under a pressure gradient of 1 atm/cm. One millidarcy (mD) is equal to 0.001 darcy (Subramanya, 2013; Ragunath, 2006 and Todd and Mays, 2005):

1.2   ASSUMPTIONS OF DARCY'S LAW

Darcy's law is a simplified mathematical relationship and as such is influenced by a number of assumptions. It is necessary to clearly state and describe these assumptions so that they are simplified and characterized mathematically.

- **Steady flow:** The primary assumption is that the flow must be in an equilibrium state or steady-state condition. It means that the law assumes that the specific discharge is not time dependent.
- **Constant temperature:** The temperature is assumed to be constant for the validity of the law as density and viscosity of fluid are related to hydraulic conductivity. This assumption is mostly valid for all groundwater systems as in groundwater; there is not a large variability in temperature.
- **Laminar flow:** The law assumes that the flow should be laminar as there is no explanation for the non-linear dissipation of energy through turbulent flow.
- **Uniform media/single fluid:** For validity of this law, hydraulic conductivity ought to be constant for the media. As such, non-uniform media cannot

be employed because it has a variable hydraulic conductivity. Only single-phase fluid is valid because it depends upon the properties of fluid.

- **Incompressible fluid:** The law is only valid when there is a linear relation between hydraulic head and elevation changes.
- **Constant cross-sectional area:** This law takes into account only constant cross-sectional area because if the cross-sectional area is variable, then the gradient will also be variable.

Darcy's flux: Darcy's flux is defined as:

$$q = \frac{v}{\varnothing} \tag{1.2}$$

where
V: Velocity of flow (m/second) and
$\varnothing$: Porosity of the porous media.

Example 1.1.1

The hydraulic gradient of an aquifer is −0.02, and hydraulic conductivity is given by 1×10^{-5} m/second. Calculate Darcy's flux for the aquifer. Assume porosity as 30%.

Solution:

Velocity is given by:

$$V = K \times i$$
$$V = 1 \times 10^{-5} \times (-0.02)$$
$$V = 2 \times 10^{-7} \text{ m/s}$$

Darcy's flux is given by:

$$q = \frac{2 \times 10^{-7}}{0.3}$$
$$q = 6.6 \times 10^{-7} \text{m/s}$$

1.3   SEEPAGE VELOCITY

Although Darcy's flux has velocity units, it is not the velocity of water in the pores. Some of the flow area is taken up by the solid matrix. The seepage velocity, V, is the average pore water velocity and is calculated as follows:

$$V = \frac{Q}{A\phi} = \frac{q}{\phi} \tag{1.3}$$

where

ϕ: Porosity of the porous media and
q: Specific discharge.

Example 1.1.2

For a sandstone aquifer, the discharge is 3.15 m³/day and the porosity is 30% with the area of flow 1 m². Calculate the seepage velocity of the aquifer.

Solution:

Specific discharge is given by:

$$q = \frac{Q}{A}$$

$$q = \frac{3.15}{1}$$

$$q = 3.15 \text{ m/day}$$

Seepage velocity is given by:

$$v = \frac{q}{\varnothing}$$

$$v = \frac{3.15}{0.3}$$

$$v = 10.5 \text{ m/day}$$

1.4　TRANSMISSIVITY

It is defined as the rate at which water moves across a unit width of aquifer under a unit hydraulic gradient. In saturated groundwater analysis with nearly horizontal flow, it is usual to combine the hydraulic conductivity and aquifer thickness into a single variable called transmissivity.

$$T = K.B \qquad\qquad (1.4)$$

where

T: Transmissivity (m³/s).

Example 1.1.3

A confined horizontal aquifer has a thickness of 15 m and a hydraulic conductivity of 20 m/day. Estimate the transmissivity of the aquifer.

Solution:

Transmissivity is given by:

$$T = K \times B$$
$$T = 20 \times 15$$
$$T = 300 \text{ m}^2/\text{day}$$

1.5 INTRINSIC PERMEABILITY

The ability of a soil or rock to transfer a fluid is defined by its permeability. When fluids other than water are at standard conditions, the permeability of the media takes the role of conductivity. The following is the relation between hydraulic conductivity and permeability:

$$k = \frac{k'\rho g}{\mu} = \frac{kg}{\vartheta} \tag{1.5}$$

where

k': Intrinsic permeability, m^2;
μ: Fluid absolute viscosity;
ϑ: Fluid kinematic viscosity, m^2/second; and
g: Acceleration due to gravity, m/s^2.

Example 1.1.4

A field test for permeability recorded a discharge velocity of 0.0417 cm/s with a hydraulic gradient of 1×10^{-2}. If $v = 0.01$ cm/s, estimate coefficient of permeability and intrinsic permeability of the aquifer.

Solution:

Coefficient of permeability is given by:

$$K = \frac{4.17 \times 10^{-2}}{1 \times 10^{-2}}$$

$$K = 4.17 \text{ cm/s}$$

Intrinsic permeability is given by:

$$K_o = \frac{K \times v}{g}$$

$$K_o = \frac{4.17 \times 0.01}{981}$$

$$K_o = 4.25 \times 10^{-5} \text{ cm}^2$$

Since 9.87×10^{-9} cm^2 = 1 darcy

$$K_o = 4307 \text{ darcy}$$

1.6 EXPERIMENTAL SETUP FOR DARCY'S LAW

An experimental setup for verifying Darcy's law is shown in Figure 1.2.

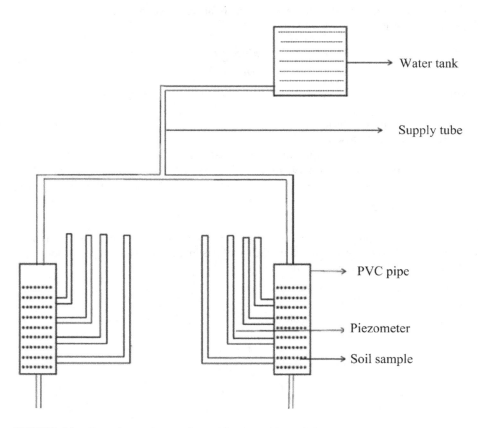

FIGURE 1.2 Experimental setup for verification of Darcy's law.

1.6.1 PROCEDURE

1. The above tank is filled with water.
2. Water is supplied to the soil column via a pipe mounted on a vertical wall. At regular intervals, a series of manometers are also attached to the soil column.
3. The head in various piezometers is measured when the soil is exposed to the atmosphere and the flow through the soil is measured at the same time.

1.6.2 OBSERVATIONS

Discharge through soil column in 5 minutes.

i. _____ L.
ii. _____ L.
iii. _____ L.
    Measured _____ L.
    Discharge of soil column is _____ L/second.
    Cross-section area of soil column _____ cm
    Length of soil column _____ cm

1.6.3 HEAD OF PIEZOMETER

i. _____ cm
ii. _____ cm
iii. _____ cm

1.6.4 CALCULATIONS

Hydraulic conductivity k in 1–2 seconds:

$h_{1-2} =$ _____
Hydraulic conductivity = _____ 2–3
Hydraulic conductivity = _____ 3–4
Mean, $k =$ _____

1.6.5 CONCLUSION

Thus, the Darcy's law is given by:

1.6.6 GUIDELINES AND PRECAUTIONS

- Using the formulas for every type of permeameter, the hydraulic conductivity must be calculated.
- Hydraulic conductivity must only be calculated from the slope of the line obtained from the plot of gradient and discharge. However, if the relation is not linear, then the points must be chosen which are closest to the origin.

- Seepage velocities and Reynolds number must be calculated.
- The values must be compared and validations and assumptions related to the law must be discussed.
- Average conductivity for normal flow must be calculated.
- All the results must be thoroughly discussed.
- Calculated values must be tabulated and presented.

Example 1.1.5

In a constant head permeameter test, the following observations were taken:
Distance between piezometer tappings = 100 mm
Difference of water levels in piezometers = 60 mm
Diameter of the test sample = 100 mm
Quantity of water collected = 350 mL
Duration of test = 270 seconds
Determine the coefficient of permeability of the soil.

Solution:

W.K.T.,

$$k = \frac{qL}{Ah}$$

In this case,

$$q = \frac{350}{270} = 1.296 \text{ mL/sec}$$

Therefore,

$$k = \frac{1.296 \times 10}{\frac{\pi}{4} \times 10^2 \times 6}$$

$$k = 0.0275 \text{ cm/sec}$$

Example 1.1.6

The difference of water levels in two observation wells at a horizontal distance of 60 m is 5 m. Determine the discharge through the aquifer per unit width if $k = 0.7$ mm/sec. The depth of the aquifer normal to the direction of flow is 2.951 m, and the hydraulic gradient is 0.082.

Solution:

From Darcy's law, discharge per unit width is given by:

$$q = k \, i \, A$$

$$q = 0.7 \times 10^{-3} \times 0.082 \times (2.95 \times 1)$$

$$q = 0.169 \times 10^{-3} \, \text{m}^3/\text{sec} = 0.169 \, \text{L/sec}$$

Example 1.1.7

A field sample of an unconfined aquifer is packed in a test cylinder. The cylinder has a length of 60 cm and diameter of 8 cm, respectively. For a period of 5 minutes, the field sample is tested under a constant head difference of 19.2 cm. It leads to a collection of 57.8 cm³ of water at the outlet. Estimate the hydraulic conductivity of the aquifer.

Solution:

The cross-sectional area of the aquifer is given by:

$$A = \frac{\pi D^2}{4}$$

$$A = \frac{\pi \times (0.08)^2}{4}$$

$$A = 0.005024 \, \text{m}^2$$

The hydraulic gradient is given by:

$$\frac{dh}{dl} = \frac{(-19.3)}{60}$$

$$\frac{dh}{dl} = -0.321$$

The average flow rate is:

$$Q = \frac{57.8}{5}$$

$$Q = 11.56 \ \text{cm}^3/\text{min}$$

$$Q = 0.0161 \ \text{m}^3/\text{day}$$

Using Darcy's law:

$$Q = -KA \frac{dh}{dl}$$

$$K = -\frac{Q}{A\left(\dfrac{dh}{dl}\right)}$$

$$K = -\frac{0.0161}{0.005024 \times (-0.321)}$$

$$K = 9.98 \text{ m/day}$$

REFERENCES

Raghunath, H.M. 2006. "Groundwater". *Hydrology: Principles, Analysis, Design*, Second Edition, New Age International Limited Publishers, New Delhi, India, 192–207.

Subramanya, K. 2013 "Groundwater". *Engineering Hydrology*, Fourth Edition, McGraw Hill Education Private Limited, New Delhi, India, 389–431.

Todd, K.D. and Mays, W.L. 2005. "Groundwater movement". *Groundwater Hydrology*, Bill Zobrist, Jennifer Welter and Valerie A. Vargas (Eds.), Third Edition, John Wiley and Sons Inc., Hoboken, NJ, 86–91.

2 Estimating Groundwater Balance and Study of Artificial Groundwater Recharges Structures

Objective: Estimating groundwater balance

2.1 PRINCIPLE

Precipitation in the form of rainfall is the most fundamental source of water. Precipitation runoff pours into streams and rivers or gathers in surface depressions to form tanks or pounds. Water from streams and rivers is held in reservoirs and then transferred directly to irrigation canals. Through appropriate conveyance systems, runoff water held in tanks or ponds is also regulated for irrigation. Groundwater is used to store a portion of the rainfall. When it rains, a portion of it intercepted on vegetation evaporates quickly. A portion of the water infiltrates into the soil, while the rest runs off onto the ground surface as runoff.

2.2 ESTIMATING GROUNDWATER BALANCE

The water balance can be explained by the concept of hydraulic equation, which is essentially a description of the law of conservation of mass applied to the hydrologic cycle. It specifies that whatever water enters a designated area ought to go into storage within its bounds, be consumed there, be exported there, or flow out on the surface or underground within a specified time period. The water balance requires that the components of supply and disposal be in balance. Over a month, season, year, or even many years, a significant difference in natural inflow and outflow is to be expected, and this difference is accounted for by the aggregate change in surface water, soil water, and groundwater storage. The unsteady state induced by climatic fluctuations would be depicted by an equation for successive short periods; the longer the period over which the hydrologic equation is drawn, the closer the several components of balance equation approach the steady state because they tend to approach an average value for climatic conditions.

DOI: 10.1201/9781003319757-3

2.3   WATER BALANCE EQUATION

The water balance or water budget is a well-known method for evaluating a basin's long-term yield. For a water table situation, the groundwater basin is assumed to be a static, three-dimensional storage box, with the following input and output:

$$R - D - Q = \Delta V \tag{2.1}$$

and

$$\Delta V = A_b S_y \Delta Z \tag{2.2}$$

where

R: Recharge to the aquifer basin;
D: Discharge or flow from the basin;
Q: Withdrawal by pumping;
ΔV: Change in volume of storage in specified time;
A_b: Area of basin;
S_y: Specific yield of aquifer; and
ΔZ: Average basin-wise change in groundwater elevation (rise or fall).

Groundwater recharge is influenced by topography, soil profile characteristics, rainfall patterns, and vegetation type. The effects of each of these variables require investigation. The water balance method has the drawback of not being able to necessarily determine the scale of potential groundwater development based on an area's water budget. The basin is treated as a surface water reservoir in the input–output evaluations. The basin's response is expected to be swift, with effects evenly dispersed throughout the basin. However, in most groundwater systems, the reactions are not evenly distributed, and water levels in a basin will fall over time as a result of depletion. To build gradients toward a well, some water must be removed from the groundwater reservoir. Two essential consequences of a long-term water supply policy are that: (i) some water must be taken from the groundwater storage in order to build groundwater supply and (ii) time delays in the areal distribution of pumping effects differ significantly from those in surface water systems. As a result, for some well systems, it is achievable. As a result, certain wells may run dry long before the groundwater system as a whole achieves a balance between replenishment and natural and outflow rates.

2.4   COMPONENTS OF GROUNDWATER RECHARGE

Groundwater resources are climatologically dynamic and expand in tandem with canal irrigation growth. The following components contribute to basin's yearly groundwater recharge.

Total annual recharge is summation of rainfall recharge, canal seepage, deep percolation from irrigated area, inflow from influent rivers, and recharge from tanks, lakes, submerged lands, etc.

The Government of India established a groundwater estimation committee in 1982, which advised that groundwater recharge be calculated using the groundwater fluctuation in all cases. However, in the absence of regular groundwater monitoring or reliable data on groundwater changes, the following ad hoc rules (Central Ground Water Board, 1986) can be used to predict groundwater recharge when planning groundwater use:

2.4.1 RECHARGE FROM RAINFALL

I. Alluvial areas
    Areas with higher clay content   10%–20% of normal rainfall.
    Sandy areas   20%–25% normal rainfall.
II. Semi-consolidated sandstones (friable
    and highly porous)   10%–15% of normal rainfall.
III. Hard rock areas
    a. Granites terrain, weathered, and
        fractured   10%–15% of normal rainfall.
        Unweathered   5%–10% of normal rainfall.
    b. Basaltic terrain, vesicular, and
        jointed basalt   10%–15% of normal rainfall.
        Weathered basalt   4%–10% of normal rainfall.
    c. Phyllites, limestones, quartzites,
        shales, etc.   3%–10% of normal rainfall.

2.4.2 RECHARGE DUE TO SEEPAGE FROM UNLINED CANALS

- For unlined canals in the standard soil with some clay content along with sand, 15–20 ha-m/day/106 sq m of wetted area of canal or 1.8–2.5 cumec/106 sq m of wetted area is recommended.
- For unlined canals in sandy soils, 25–30 ha-m/day/106 sq m of wetted area or 3–3.5 cumec/106 sq m of wetted area is recommended.
- For lined canals, seepage losses can be calculated as 20%.

Example

Bhavani basin is the fourth largest subbasin in the Cauvery basin. The entire command area of all three major canals that takes off from the Bhavani river falls within the Erode District, that is, Lower Bhavani Project (LBP), Kodiveri, and Kalingarayan canals. The LBP canal is a major source of irrigation in Erode District. Many of these canals are unlined and leakage takes place from them. Thus, the seepage from the canal helps in recharging the wells in the area. LBP canal is designed as an unlined canal. Due to the seepage loss, groundwater recharge is 39.54%.

2.4.3 Return Seepage (Deep Percolation) from Irrigated Fields

I. Irrigation using surface water sources
a. One-third of water that is delivered is actually used in the field and the disparity in the percentage of seepage may be governed by studies undertaken in the area or in a similar area.
b. 40% of water is distributed to paddy irrigation outputs.
II. Groundwater irrigation accounts for 30% of the water delivered at the exit. For paddy irrigation, up to 35% of the provided water may be used as returned seepage.

Return seepage figures in all of the aforementioned examples include losses in field channels, which should not be individually accounted for.

Example

A paramount source that can be beneficially used for recharging shallow groundwater aquifers is seepage from irrigation canals. It has been revealed that groundwater rises to a great extent below the canals by recharge from irrigation canals.

2.4.4 Seepage from Tanks and Ponds

Tank seepage typically ranges from 9% to 20% of their live storage capacity. Although data on the live storage capacity of a significant number of tanks may not be accessible, seepage from the tanks could be calculated at 44–60 cm per year across the overall water spread area.

The seepage from the percolation tank is higher, accounting for up to half of the total storage capacity.

When it comes to seepage from ponds and lakes, the same rules that apply to tanks can be applied. (The recharge component from percolation tanks, as estimated, shall be distributed solely for the purpose of consumption under its direction.)

Example

There are around 500,000 tanks and ponds located all over India, especially in peninsular regions. The purpose of these tanks is irrigation, but the command area of these tanks has numerous wells located in its periphery, which get recharged due to the seepage from these tanks and ponds.

2.4.5 Contribution from Influent Seepage

The influent seepage from rivers with a definite influent nature may be estimated by using Darcy's law:

$$Q = T i L \tag{2.3}$$

where

Q: Rate of flow, m³/day;
T: Transmissibility of the aquifer, m²/day;
i: Hydraulic gradient, m/km; and
L: Length of river section through which flow is taking place, km.

2.5 GROUNDWATER LOSSES

The loss of groundwater from aquifers occurs mainly from the following:

1. Outflow to rivers.
2. Transpiration by trees.
3. Evaporation from water table.

2.5.1 RIVER OUTFLOW

Depending on the relative water levels in the river and adjacent aquifers, a river may have a losing or growing reach. The effect of losing reach on the groundwater regime is cumulative.

2.5.2 TREE TRANSPIRATION

Tree transpiration leads to the depletion of groundwater equivalent to the amount of water required for tree growth. This is a groundwater loss that must be viewed as irreversible.

2.5.3 EVAPORATION FROM A SHALLOW WATER TABLE

Once the water table drops roughly 3 m below the ground level, these losses become minimal. As a result, these losses are frequently overlooked in groundwater estimates.

The total groundwater resources computed for irrigation are regenerated groundwater resources that are occasionally committed for lift irrigation schemes and other surface irrigation operations. Thus, it is considered that 15% of the total groundwater resources be maintained for drinking and industrial purposes, as well as for committed base flow and unrecoverable losses.

2.6 GROUNDWATER REPLENISHMENT

Precipitation is the source of groundwater. The rate at which the groundwater reservoir is replenished varies depending on the precipitation pattern. It also varies with the inherent permeability of the soil and other earth components through which rain falling directly on the land surface replenishes the water reservoirs. Direct seepage from surface water sources, such as streams, rivers, canals, and lakes, replenishes places underneath them, provided the intervening layer is permeable.

2.6.1 Artificial Groundwater Recharge

The long-term use of groundwater necessitates optimal development and appropriate management procedures. Average pumpage may equal or surpass average recharge in water-bearing rocks that are easily replenished from the surface. Pumpage may exceed recharge in deep artesian aquifers where a significant volume of recharge occurs in a region other than the area where it is utilized. However, in such instances, groundwater utilization must be based on quantitative estimations of how much of the total groundwater storage is consumed. However, in coastal areas, average pumping must be far lower than normal recharge to allow groundwater to flow seaward and avoid seawater from being injected into wells.

2.6.2 Methods of Artificial Recharge

Soil conservation and artificial recharge processes can help promote groundwater recharge. The amount of rainfall or ponded surface water penetrating into the soil is highly dependent on the soil surface condition and the moisture content at the time of rainfall.

Surface spreading, subsurface approaches, and indirect methods are all direct means of groundwater recharge.

2.6.2.1 Surface Spreading

The penetration of water from the surface of the soil through the vadose zone to the saturated area of the aquifer is referred to as groundwater recharge by surface spreading. The vertical hydraulic conductivity of the vadose zone, the presence or absence of limiting layers with low hydraulic conductivity, and the influence of physical, chemical, and bacteriological processes all influence the downward movement of water.

The following are the requirements for a successful recharge using the spreading method:

- The aquifer should be permeable, unconfined, and sufficiently thick.
- Surface soil should be permeable to sustain high infiltration rate.
- Permeable unsaturated zone which is free of clay lenses.
- The groundwater level should be ample deep in order to make room for the rise in the water table without causing waterlogging.
- The aquifer material should have adequate hydraulic conductivity.
- Land should have a gentle slope with no gullies or ridges.

2.6.2.2 Subsurface Techniques

Because the confined aquifer cannot be recharged by infiltration from the soil surface due to the intervening impermeable layers between unconfined and confined aquifers, subsurface techniques are utilized to recharge it. The following are some of the most commonly used subsurface techniques:

- Injection wells
- Recharge wells for gravity heads

- Connector wells
- Recharge pits
- Recharge shaft

2.6.2.2.1 *Injection Wells*

Injection wells are basically tube wells which are pumped with treated surface water to rehydrate a confined aquifer (Figure 2.1). This approach is useful when space is limited, such as in metropolitan areas, and it can also be used in coastal areas to prevent seawater penetration. The wells are built like gravel-packed pumping wells that tap a single or numerous aquifers. The only dissimilarity is that the well's upper part is filled with cement to prevent pressure leakage from the borehole and well assembly's annular space. The injection wells become clogged with time. The recharge water should be adequately treated to remove suspended particles and hazardous microorganisms to avoid blockage.

Example

The artificial recharge using injection wells has been well carried out in the Ghaggar River basin in Punjab. For this purpose, canal water is used as the primary surface

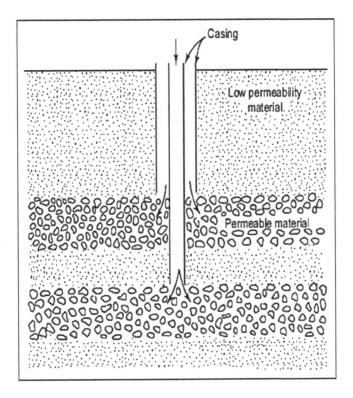

FIGURE 2.1   Injection well technique.

water source. Normally, the injection rate initially is 43.80 L/second, having an injection pressure of 1 atmosphere (atm). The pressure then escalates to 2 atm after about 5 hours and then remains sustained thereafter, albeit the recharge rate gradually plummets to 3.5 L/second after few days. The natural, gravity-controlled recharge rate is 5.1 L/second.

2.6.2.2.2 Gravity Head Recharge Wells

Surface water can be directed to bore wells, tube wells, and dug wells to recharge groundwater. The aquifer is replenished by gravity. The relief between the surface water level in the feeder reservoir and the elevation of the water table, or piezometric head, is the recharge head available. When compared to recharge via injection wells, the rate of recharge is significantly lower. In comparison with poor-yielding wells, abandoned wells that were producing a decent yield prove to be more suitable for groundwater recharge. For recharging, only well-filtered and disinfected water should be used. To minimize impact waves, airlock, and other problems, the water should be channeled to the well through a conductor pipe.

2.6.2.2.3 Connector Wells

Connector wells are used to recharge groundwater from one aquifer to the next. Aquifers with higher heads begin to recharge aquifers with lower heads. Deeper aquifers are typically refilled from phreatic aquifers. As a result, the connector effectively joins all of the aquifer zones observed during the boring process.

Example

The artificial recharge in the Central Mehsana area of North Gujarat is carried out utilizing injection wells, connector wells along with infiltration channels and ponds.

2.6.2.2.4 Recharge Pits

Recharge pits are essentially used when phreatic aquifers are not hydraulically connected to surface water because of the presence of impermeable low/permeable layers or lenses which hamper the penetration of surface water into the aquifer. The recharge pits (Figure 2.2) recharge through the unsaturated zone and do not penetrate or reach the unconfined aquifer as a gravity head recharge well must. The water used for recharging should be free of sediment as much as feasible. To avoid clogging of permeable strata, a thin layer of filter should be placed at the bottom of the pit.

Example

Artificial recharge in the coastal areas of Saurashtra is implemented using recharge basins. The release of water from the canal system of Hiran Irrigation Project is used as the water source for this purpose.

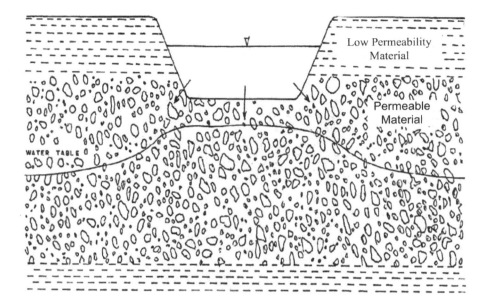

FIGURE 2.2   Sectional view of recharge pit.

2.6.2.2.5   Recharge Shaft

The recharge shaft and recharge pit are almost similar in appearance, but the former has a smaller diameter (Figure 2.3). It is appropriate when the water table aquifer is overlain by strata that are less permeable. The recharge shafts can be manually excavated at shallow depths and non-caving strata, but they can also be drilled by a reverse/direct rotary drilling rig for greater depths. The diameter can range between 0.8 m (drilled shaft) and 2.0 m (manually excavated). The contrast between a recharge well and a shaft is that the shaft ends in recharge shafts have more pervious strata beneath the confining layer but do not reach the water table, while on the contrary, recharge wells penetrate the entire confined aquifer. The shallow shafts are replenished with stones from an inverted filter, and the top few meters with gravel and sand from a gravel and sand filter.

2.6.2.3   Indirect Methods

The indirect methods include:

   i. Induced recharge;
  ii. Aquifer modification; and
 iii. Groundwater conservation techniques.

2.6.2.3.1   Induced Recharge

Induced recharge invariably means to induce artificial recharge from surface water which is hydraulically connected to the aquifer. This can be accomplished by pumping groundwater through wells, collector wells, and infiltration galleries, depending on the geo-hydrological condition of the area.

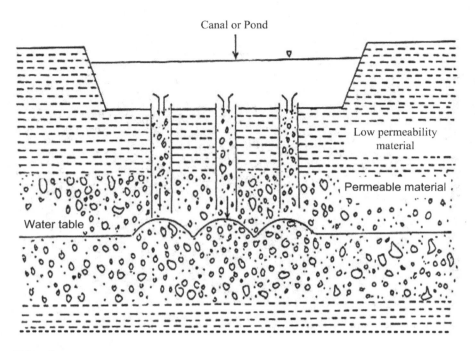

FIGURE 2.3   Sectional view of groundwater recharge shaft at the bottom of a canal.

2.6.2.3.2   Aquifer Modification

Aquifer alteration techniques such as bore well blasting and hydro-fracturing change the aquifer's characteristics, increasing its potential for groundwater recharge. As a result, rather than artificial recharge, these strategies are artificial yield augmentation measures. Furthermore, these strategies are employed not only to raise the productivity of producing wells but also to increase inflow in the surrounding areas. In regions where there is a lot of hard rock, several procedures are applied.

2.6.2.3.3   Groundwater Conservation Techniques

Groundwater dams/underground *Bandharas* constitute the most prevalent groundwater conservation technology (Figure 2.4). A subsurface barrier is built across a nala (seasonal torrent) bed, stream, micro-watershed, or other area in this manner to hold water flowing below the ground surface in its upstream direction. As a result, the groundwater leaving the basin is halted and stored within the aquifer. The following are the requirements for using this technique:

- The stream is connected to a phreatic aquifer hydraulically;
- The valley is clearly defined, wide, and has a restricted exit (bottlenecked);
- The unconfined aquifer is found at a depth of 10–20 m below the ground level; and
- The aquifer beneath the site has a sufficient thickness (minimum 5 m).

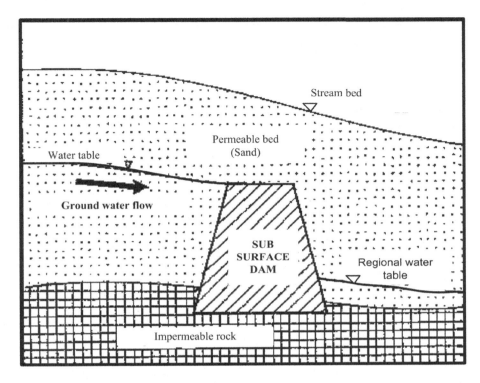

FIGURE 2.4 Schematic view of subsurface dam.

REFERENCES

Kumar, C. P. and Seethapathi, P. V. 1988. "Effect of Additional Surface Irrigation Supply on Ground Water Regime in Upper Ganga Canal Command Area, Part I - Ground Water Balance", Case Study Report No. CS-10 (Secret/Restricted), National Institute of Hydrology, Roorkee, 1987–1988.

Michael, A.M., Khepar, S.D. and Sondhi, S.K. 2008. "Groundwater resource development and utilization". *Water Wells and Pumps*. Tata McGraw-Hill Publishing Company Limited, New Delhi, India, 1–52.

Ranghavan, K. 1967. Influence of tropical storms on monsoon rainfall in India. *Weather*, vol. 22, pp. 250–256.

Sengupta, N. 2006. Fragmented landholding, productivity, and resilience management, *Environ. Dev. Econ.*, vol. 11, no. 4, pp. 507–532.

Simmers, I. 1988. *Estimation of Natural Groundwater Recharge*, Springer, Dordrecht, https://doi.org/10.1007/978-94-015-7780-9

Singh, P.K., Dey, P., Jain, S.K. and Mujumdar, P. 2020. Hydrology and water resource management in ancient India, *Hydrol. Earth Syst. Sci.*, vol. 24, no. 10, pp. 4691–4707.

Sinha, B.C. and Sharma, S.K. 1988. *Natural Groundwater Recharge Methodologies in India*, Springer-Nature, Netherlands. 301–311.

Sophocleous M. 1987. Basinwide water-balance modeling with emphasis on spatial distribution of ground water recharge, *J. Am. Water Resour. Assoc.*, vol. 23, no. 6, pp. 997–1010.

3 Water Table Contour Maps

Objective: Draw Water Table Contour Maps

3.1 THEORY

The water table is the top of the saturated zone, below which water fills all soil pores and rock fissures. Groundwater travels through the subsurface in the same way as surface water does, although at a much slower rate. Groundwater can travel up to 5 ft/day if the soil is primarily sand and gravel. Groundwater, on the other hand, usually moves at a rate of a few inches each day (or less). Groundwater, for instance, streams and rivers, travels from higher to lower elevations. To illustrate how groundwater moves below the ground, down the sides of a valley, and into a river that flows to the sea, the contours of the water table will be drawn. At the point where the water table crosses the land's surface, groundwater is discharged to surface waterways. The surface water in this circumstance is known as a gaining stream or gaining pond. As the water table is adjacent to "gaining" surface waters, the elevation of groundwater is usually the same as that of the river, chiefly between rainstorms. Groundwater is required to flow perpendicular to the contours of the water table. This happens due to the fact that groundwater has a tendency to move downhill in the path of least resistance because of gravity. All three of these ideas will be applied in this exercise. Drawing a water table contour map will be demonstrated in this assignment. Water table estimations that are taken at the corresponding time of the year can be used to draw a water table contour map that portrays the groundwater flow direction. Monitoring wells are commonly used to determine the water table's elevation which is measured at many points throughout the study region. Water table contours, like topographic map contours, indicate the lines of equal elevation. The elevations of water table are quantified in wells and at the river channel but not on the ground surface, as presented on the maps (Figure 3.1). As a result, like surface water flows downward and at right angles to topographic contours, groundwater moves downhill and at right angles to water table contours as well (Predicting groundwater flow, 2018).

3.2 EXPERIMENTAL PROCEDURE

1. Using the "Contouring the Water Table" worksheet (Figure 3.1), lightly draw three or four arrows to represent the prediction for the direction(s) of groundwater flow (in case one makes any mistakes).
2. Draw contour lines every 50 foot. It is always possible to ink over the pencil lines later. Begin with a 50-foot contour (the ocean's shoreline will be at sea level), and then move up to 100, 150, 200, and 250 ft.

DOI: 10.1201/9781003319757-4

Contouring the Water Table

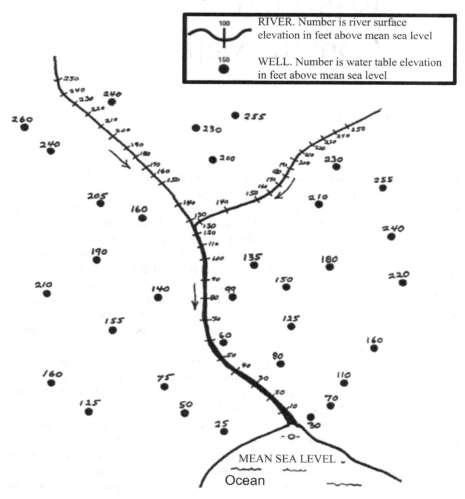

FIGURE 3.1 Contouring the water table.

3. Begin by drawing a 50-foot contour. On the river, look for the 50-foot elevation. A line should be drawn from that point to the well just southwest of the river, which has a 50-foot elevation.

4. A contour should be drawn on other side of the river. One must interpolate—that is, find out the proportional distance between two points—when locating a contour between two points. Between the 30- and 80-foot altitudes, the 50-foot contour should be drawn nearer to the 30-foot value (20-foot difference) than the 80-foot value (30-foot difference). After a little practice, one can do it by hand or use a ruler and calculator to measure it

precisely. Because the 30- and 70-foot elevations are both 20 ft different from the 50-foot contour's value, draw the contour directly between them for the other two wells.

5. When one is done, one will note that the river and its tributaries create versus the contours. Because the river is a "gaining" river, this is the case. It is being refilled by the aquifer. Groundwater is moving down the valley's sides and into the river channel, as shown by the contours. A "losing" stream is the polar opposite of a "gaining" stream. This phenomenon usually takes place due to the water table at the stream channel being lower than the elevation and stream water flows downhill across the channel to the water table.

Note: It is critical to have precise well and stream elevations when generating a water table map. All heights should be compared to a common reference point, for example, mean sea level. All altitudes are either above or below the standard datum in this case (e.g., 50 ft above mean sea level datum). It is also critical to take a "snapshot" of all water table elevations over a short period of time, such as 1 day, so one can see what is going on. Due to the rise and fall of water table over time, taking readings before these changes occur helps ensure that one's map is more accurate.

3.3 INTERPRETATION

- Subsurface geology plays an important role for the actual interpretation of water table contour map in addition to the consideration of topography, recharge, discharge patterns, and natural drainage.
- It is important to take into account the spatial distribution of permeable layers that are present below the water table.
- The hydraulic gradient of the water table is literally represented graphically using the water table contour maps.
- Darcy's law controls how well the hydraulic gradient of water table is represented by water table contour maps (Karin, 2018).

3.4 GUIDELINES

- When values take into account surveyed datum, then all water elevations must be measured from mean sea level.
- For the movement of contaminant flumes as well as ascertaining the direction of groundwater flow, water table maps should be used.
- When more than three points are used in water table maps, abutting triangular areas must be used.
- Elevation points lying along a line make it difficult to draw a water table map. One of the most convenient methods to make water table maps is using three points that lie in a triangle (Karin, 2018).

3.5 PRECAUTIONS

- A professional hydro-geologist must be consulted to appropriately interpret the data which are utilized in decision-making as a number of assumptions can influence the making of water table maps.
- There is uncertainty in the accuracy of contoured map when elevation points lie at a distance from one and thus a greater uncertainty lies in using the map for resource decisions.

3.6 GROUNDWATER CONTOUR MAP EXAMPLE

Table 3.1 contains the well-gauging data. Nine monitoring wells were used to determine the depth to groundwater. At the uppermost part of each monitoring well, the ground elevation was measured. For calculating the elevation of the water table, the difference between the depth to groundwater from the elevation of the top of the well is estimated.

Example:

Top of well elevation = 84.30 ft
Depth to groundwater = 7.02 ft
Elevation of water table = 84.30 − 7.02 = 77.28 ft

1. The height of the water table in each monitoring well is ascertained.
2. Fill in the groundwater elevations at each well on one copy of the site map (write in pen). Also provide the elevation of the lake's surface.

TABLE 3.1
Groundwater Gauging Data

Well	Well Elevation (ft)	Depth to Groundwater (ft)	Water Table Elevation (ft)
W-1	85.20	6.04	77.28
W-2	96.75	20.21	77.66
W-3	101.45	24.75	78.14
W-4	97.21	20.37	77.54
W-5	103.71	26.34	78.84
W-6	83.63	18.73	77.38
W-7	80.89	4.34	77.21
W-10	82.33	8.23	77.23
W-11	84.32	10.33	77.13
W-12	96.55	25.28	77.43
W-13	101.88	20.45	77.95
W-14	83.96	2.58	77.19

3. The groundwater elevations are contoured using a 0.5-foot contour interval using a pencil. Each contour line should be labeled.
4. Draw an arrow indicating the direction of groundwater flow (right angles to the contour lines and downside the slope of the groundwater table).

REFERENCES

Karin, B. 2018. Kirk at Skidmore College, and Scott K. Johnson at the Fractal Planet. http://www.skidmore.edu/~jthomas/fairlysimpleexercises/gwasessment.html

Predicting Ground Water Flow. 2018. New England's Ground Water Resources. https://www.epa.gov/sites/default/files/2015-08/documents/mgwc-gwb10.pdf

Part 2

Measurement of Soil Properties

4 Sieve Analysis for Gravel and Well Screens Design

Objective: Sieve Analysis for Gravel and Well Screens Design

4.1 DEFINITIONS OF RELATED TERMS

4.1.1 GRAVEL PACKING

A gravel pack is a thin layer of coarse material (specifically gravel) produced surrounding a well's screened section. Wells can be gravel-packed naturally or mechanically (Figure 4.1). Surging and bailing can be used to transfer fine sand and silt from the natural formation through the well screen apertures, resulting in a naturally created envelope. Natural gravel packing is a term used to describe the process. The well bore should be somewhat larger than the well screen such that the screen is centered in the hole, and the annular area around the screen should be filled carefully by selecting gravel to suit the aquifer gradation. This leads to creation of an artificial gravel envelope. Artificial gravel packs are also known as "gravel shrouding" or "gravel filtering".

4.1.2 ARTIFICIAL AQUIFER AND GRAVEL PARK SAMPLES

The properties of the water-bearing formation are revealed by sieve examination of samples acquired during the drilling of test holes or production wells. The findings serve as a foundation for making judgments about well screen specification and gravel pack design.

The dry sieving method is used to analyze sand/soil samples, which is the usual procedure. The conventional procedure is to employ a set of sieves that adhere to the Indian Standard (IS: 460–1962) or the country's specific standard. Each sieve's weight of material retained is recorded. These weights are then reported as percentages of the total weight of the sample, and a graph of the cumulative percent of the sample retained on a specific sieve and all the sieves above it versus the size of the given sieve, in millimeters, is shown. On the y-axis, the "percent retained" is plotted, and on the x-axis, the size of the sieve aperture, or "particle size", is plotted. The diameter of the smallest particle retained by each sieve is used to determine the size of opening.

Figure 4.2 shows typical sand analysis curves for various aquifer materials, illustrating the formulation's water-holding capabilities and the need for artificial gravel packing.

DOI: 10.1201/9781003319757-6

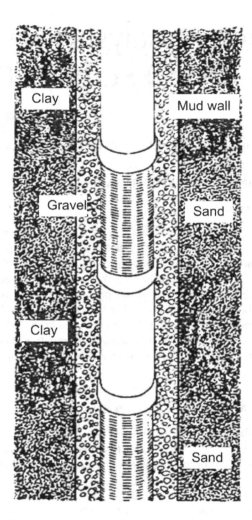

FIGURE 4.1 An artificially gravel-packed tube well showing location of slotted pipes.

4.1.3 Effective Size (D_{90})

The phrase effective size refers to the particle size distribution in which 10% of the sand is finer and 90% is coarser. Class C curve in Figure 4.3, for example, illustrates that 90% of the sample is made up of sand grains larger than 0.25 mm, whereas 10% is smaller. As a result, the formation material's effective size is 0.25 mm.

4.1.4 Uniformity Coefficient (C_U)

This is a granular material's grain size variation expressed as a ratio. It is commonly determined by the sieve aperture that allows 10% of the material to pass through.

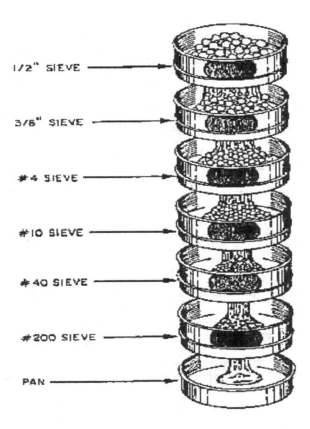

1/2" SIEVE

3/8" SIEVE

#4 SIEVE

#10 SIEVE

#40 SIEVE

#200 SIEVE

PAN

FIGURE 4.2   Set of sieves for analyzing aquifer samples.

$$C_u = \frac{D_{60}}{D_{10}}$$

As a result, the homogeneity coefficient in the class C curve (Figure 4.3) is 0.75 mm divided by 0.25 mm or 3.0 mm. Hazen (at the end of the nineteenth century) introduced this ratio as a quantitative expression of the degree of water-bearing sand assortment as a porosity indicator. The coefficient has a value of unity for complete assemblage (one grain size), two to three for generally uniform-grained sand, and a high value for heterogeneous sand.

4.1.5   FORMATIONS REQUIRING ARTIFICIAL GRAVEL PACK

Artificial gravel packing is not necessary for all water-bearing strata. Generally, formations with an effective size of 0.25 mm and a uniformity coefficient of 2 or higher can be formed without a gravel pack as long as the formation has minimal vertical sizing variations. The desirability of the gravel pack reduces as the formation becomes coarser; nevertheless, there are numerous exceptions to this rule.

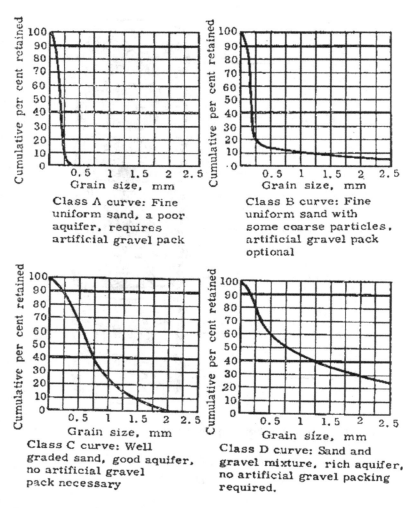

FIGURE 4.3 Typical sieve analysis curves of water-bearing sands and gravels (reference).

Where the natural formation comprises fine homogeneous sands and/or the formation is heavily laminated, artificial gravel pack building is advised which is an amalgamation of alternating fine, medium, or coarse layers that are thin and hence strenuous to locate accurately. In most deep tube wells, these conditions are most usually satisfied. When bigger screen holes are desired than those indicated by sieve analysis, an artificial gravel pack can be utilized for an aquifer containing fine particulates. When C_u is between 2.0 and 3.0 and the D_{40} size is less than 0.42 mm, this happens frequently. Large screen holes are preferable in places where encrustation is an issue.

The following are some of the benefits of gravel pack wells:

1. Gravel packing enhances the well's effective diameter and consequently its specific yield.
2. It improves a well's yield by lowering flow resistance at the well screen.
3. It lessens incrustations. Because of the large screen holes, this is the case.
4. Gravel packs, when appropriately built, give sand-free water, enhancing efficiency.
5. It gives the well screen more structural strength.
6. It prevents formation material from collapsing in, lowering the risk of clogging the well screen.
7. It makes it easier to remove the well casing and screens in shallow wells.

4.1.6 GRAVEL PACK MATERIAL

Clean, well-rounded, smooth, homogeneous grains characterize good gravel pack material (Figure 4.4).

Formation Coarse Sand   **Gravel Pack- Fine Gravel**

Formation Medium Sand   **Gravel Pack- Very Coarse Sand**

FIGURE 4.4   Relation between particle size distribution of aquifer material and gravel pack.

(These properties improve the pack material's permeability and porosity.) It should be a hard, insoluble siliceous substance with a calcareous particle content of less than 5% (limestone). Shale, anhydrite, and gypsum particles are all undesirable. The material should have a consistent size distribution, with little variation in grain sizes. The uniformity coefficient (C_u) of a material must be less than 2.0 to be considered uniform.

The uniform-grain-size pack and the graded grain size pack are the two most common types of gravel packing. The former has gained popularity in recent years; particularly when manufactured screens are utilized and the aperture size can be regulated. In the case of a graded pack, formation material may intrude at the gravel-formation contact, partially filling the pores and reducing permeability. The particles of the formation can migrate between the grains and be drawn into the well during development with a well-stored (homogeneous) gravel pack, enhancing formation permeability while maintaining the pack's highly permeable character.

4.2 DESIGN OF GRAVEL PACK

In many circumstances, the principal disadvantage of using homogeneous pack material is lack of supply. The particle size as represented by the mean grain diameter, which is 50% grain size, is the most essential physical attribute of uniform-grain-size material. The grain size of the pack does not have to be huge. The maximum particle size in the gravel pack, according to the American Society of Agriculture Engineers, should be 6.4 mm (Figure 4.5).

The other curve represents the commonly used gravel pack, which is defined as very coarse sand to fine gravel. The size ratio of gravel pack to formation is 4.8 because the mean grain size of the formation material is 0.38 mm and that of the gravel pack material is 1.8 mm. The gravel pack ratio, also known as the pack-aquifer ratio, is a ratio of mean sizes (P-A ratio).

$$P\text{-}A = \frac{50\,\%\ \text{size of gravel pack}}{50\,\%\ \text{size of aquifer}}$$

The pack-aquifer ratio is defined by some authorities as the ratio of the pack's 70% size to the aquifer's size. In most cases, the 50% threshold is used.

The lower P-A ratio should be 4.0 to reduce head loss through the gravel pack. A P-A ratio greater than 9.0 may allow sand to move, and thus this value of 9.0 may be considered a practical maximum limit. For the effective design of wells with gravel packing, Smith (1954) found that ratios of 4–5 were satisfactory. Due to sand pumping, however, wells with gravel pack ratios of 7–10 were determined to be insufficient. Smith went on to say that even greater gravel pack ratios (10–20) result in excessive sand pumping. For limiting pack-aquifer ratio as stated in Table 4.1, steady filtering action in tube wells was proposed based on model experiments undertaken at the Irrigation Research Institute, Roorkee (Gupta, 1964).

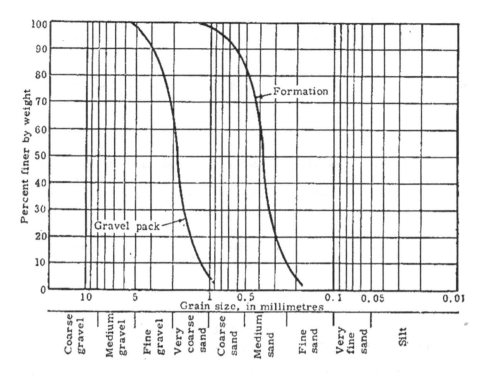

FIGURE 4.5   Sample sieve analysis curve of aquifer material and gravel pack.

TABLE 4.1
Recommended Pack-Aquifer Ratios for Stable Filtering Action in Tube Wells

Sr. No.	Type of Material		P-A Ratio—50% Basis		
	Aquifer Type	Gravel Pack	Range within Which Both Sand Movement and Resistance Are Minimum	Maximum Upper Limit after Which the Visible Failure Takes Place	Recommended Range of Stable P-A Ratio
(a)	Uniform	Uniform	9–13	29	9–13
(b)	Non-uniform	Uniform	11–16	33	11–16
(c)	Uniform	Non-uniform	12–18	35	12–18
(d)	Non-uniform	Non-uniform	15–22	42	15–22

Source: Gupta, S.N. (1964).

According to the findings of module studies done at Ludhiana (PAU, 1971), the following top limits of P-A ratios must be maintained in order to maintain a stable filtering action:

Aquifer	Gravel	Limiting P-A Ratio
Uniform	Uniform	6
Non-uniform	Uniform	9

It was also noted that slotted pipe is readily available in India in two slot sizes: 1–6 and 3.2 mm. As a result, if the size of the aquifer material necessitates slot sizes smaller than 1.6 mm, the following P-A ratio may be employed to solve the practical challenge of slot sizes smaller than 1.6 mm not being available:

Sand	Aquifer	Limiting P-A Ratio
Uniform	Uniform	8
Non-uniform	Uniform	12

4.3   DESIGN PROCEDURE OF GRAVEL PACK AND SLOT SIZE OF WELL SCREEN

The gravel pack and size of the well screen to be utilized in a well are designed using the following methods (Johnson Inc., 1966):

1. Determine the particle size distribution of different formations encountered by the well by preparing sieve analysis curves of the material retrieved from the well log.
2. Determine the formation depth where the well screen will be installed.
3. To estimate the size of the gravel to be used as the pack material, take 70% of the diameter of the specified formation and multiply it by a constant whose value varies from 4 to 6. When the formation is made up of fine particles of the same size, the multiplier 4 is employed, and when it is made up of coarse particles of varying sizes, the multiplier 6 is used. The intermediate values have been set to suit the formation material's intermediate size range. If the formation materials are non-uniformly sized sand and silt, the constant values may range from 6 to 9, depending on the degree of non-uniformity. Locate the gravel size on the particle size distribution graph, as calculated earlier (Figure 4.6). Draw a curve parallel to the particle size distribution curve of the formation across this location.

   By trial and error method, the curve should be created in such a way that it represents a uniformity coefficient of 2.5 or less. (Gravel with a homogeneity coefficient of less than 2.5 will segregate the gravel particles during placement, resulting in inefficient well operating.) The size distribution of the graves to be used for gravel packing will be represented by the

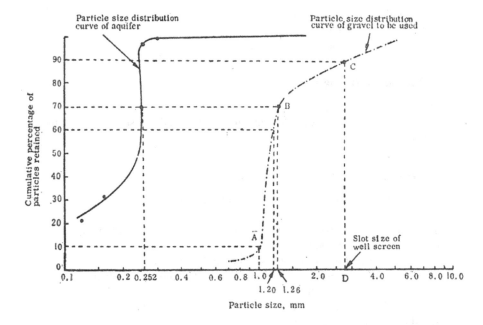

FIGURE 4.6   Design of gravel size and slot size opening for gravel pack tube well.

curve so created. This graph should more or less match the particle size distribution curve of the gravel pack used in the well.

4. Selection of size of slot opening in the screen: The particle size distribution curve of the gravel should be used to estimate the slot size opening of the well screen. Find a point C on the gravel particle size distribution curve that indicates that the gravel should be kept at 90% of its original size (Figure 4.6). Draw a line parallel to the y-axis through C, meeting the x-axis at point D, to represent the well screen's slot size. Depending on the size of the tool used to make the slots in the well screen, the actual size of the slot is fixed at ±8% of the aforementioned size.

4.4   LABORATORY PROCEDURE

4.4.1   SCOPE

The method is specifically employed for determining the particle size of various aggregates using the sieve analysis test. A procedure for accelerating the testing for provision of accurate results has been described as follows.

4.4.2   APPARATUS REQUIRED

a. A balance scale for weighing the samples,
b. Woven-wire sieves,

c. Sieve shaker,
d. Mechanical washing vessel, and
e. An oven or preferably heating device.

4.4.3 MATERIALS REQUIRED

- Distilled water for washing the sample.
- A wetting agent, for example, calgon, can be used for separation of fine particles.

4.4.4 PROCEDURE FOR SIEVE ANALYSIS TEST

- Dry soil sample should be weighed and the weight must not be less than 500 g.
- The weight of the sieves as well as the pan that will be utilized in the experiment should be recorded.
- The sieves should be thoroughly cleaned before the test.
- The sieves must be placed in ascending order and the ones with larger openings must be placed on the top.
- This implies that the number 4 sieve should be at the top while number 200 should be at the bottom of the stack.
- The soil sample should be placed on the top sieve and covered with a lid.
- The stack is to be placed in a mechanical shaker and shaken for about 10 minutes.
- After 10 minutes, the sieve stack is removed from the shaker.
- Weights of each sieve as well as the pan at the bottom of the stack are measured.

4.4.5 CALCULATIONS

- The grading of the samples should be calculated.

4.4.6 INTERPRETATION OF DATA

- Calculating the difference between the weight of empty sieve and the recorded weight of sieve after performing the test gives the weight of soil retained on each sieve.
- The initial weight of the soil sample is compared to the summation of the total weight of particles retained on the sieve. The difference should be lower than 2%.
- The percentage of particles retained on each sieve is calculated by dividing each weight retained by the initial weight of the soil sample.
- fterward, the total percentage passing from each sieve is calculated by finding the difference between the cumulative percentage retained in that particular sieve and the ones above it from total.

4.4.7 Precautions

- Sieves should never be overloaded. At the completion of test, the material retained on the sieve should not be more than two layers deep as more depth will result in blinding of the mesh and hence imprecise results.
- For total separation, enough time should be allowed on a sieve shaker. The difference in the percent passing should never exceed 1%. If the difference is more than 1%, then the process should be repeated until the target of 1% is achieved.
- Degradation should be checked properly. Fragile and crumbly materials have a tendency to break apart during separation which leads to imprecise results.
- Shaker time intervals should be reduced depending on the conditions.
- Sieves should be checked consistently for defects like wear or distorted openings as it can lead to prejudiced results.
- Tools or probes should never be used to remove trapped particles as it will result in higher chances of mesh damage.
- Absolute care should be taken while transferring the material to the tare weighing pan.
- Sample used for the analysis should always be pre-dried.

Example 4.1

Table 4.2 shows the size distribution of the aquifer material from 10 to 12 m from the ground surface of a well penetrating a semi-confined aquifer in an alluvial terrain. Design the size of gravel pack and slot size opening of a slotted pipe well screen.

TABLE 4.2
Sieve Analysis of Lithological Section between 10 and 12 m in Example 4.1

Size of the Sieve Opening (mm)	Cumulative Weight Retained (g)	Cumulative Percentage Retained	Cumulative Percentage Finer
>2.54	0.00	0.00	100.00
1.80	5.00	1.00	99.00
0.30	16.00	3.20	95.00
0.25	318.00	63.60	32.20
0.21	3.00	0.60	31.60
0.16	50.00	10.00	21.60
0.12	38.00	7.60	14.00
<0.12	70.00	14.00	0.00

Solution

Curve A in Figure 2.1.6 depicts the particle size versus the proportion of formation material preserved at a depth of 10–12 m from the ground surface. On the curve, the 70% diameter of the forming material is 0.252 mm. A multiplying factor of five is employed to achieve a uniform size of gravel, resulting in a size of 1.26 mm (0.252 × 5). B is the location of the point. A curve is constructed through point B that is roughly parallel to the formation's particle size distribution curve and represents a uniform coefficient less than 2.5. The gravel's D_{60} size is 1.2 mm and the D_{10} size is 1 mm. This results in a homogeneity coefficient of 1.2, which is considerably below Johnson Inc.'s recommended limit of 2.5 (1996). The gravel has a diameter of 3.01 mm at 90%. As a result, the well screen's slot size is set to 3 mm.

4.5   GRAVEL PACK THICKNESS

Because the mechanical retention of formation particles is the basis for gravel pack gradation design, a pack thickness of only two or three grain diameters is all that is required to maintain and manage the formation of sand. Edward E. Johnson, Inc. (1996) conducted laboratory tests and found that a gravel pack with a thickness of only a fraction of a centimeter successfully maintains formation particles despite the velocity of water tending to move the particles through the gravel pack. However, placing a fraction of a centimeter thick gravel pack in a well and expecting the material to entirely encircle the well screen are unrealistic. A minimum thickness of 7.5 cm is regarded practical for field installation to ensure that an envelope of gravel surrounds the entire screen. The upper limit of gravel pack thickness should be around 20 cm in most cases. A thicker envelope has no discernible effect on the well's yield. Because the regulating factor is the ratio of the grain size of the pack material to the grain size of the formation material, thickness does nothing to limit the potential of sand pumping. A gravel layer that is too thick can make the well's final development more challenging.

Example 4.2

The results of sieve analysis test carried out on a 400 g sample of underground aquifer proposed to be tapped for installation of a tube well at a location are given in the following tables (Figure 4.7):

Sieve Sizes (mm)	Mass Retained (g)
4.75	17.88
2	3.91
1.18	5.03
0.425	112.15
0.3	76.52
0.15	67.53
0.075	12.5
<0.075	104.48

Sieve Sizes (mm)	Mass Retained (g)	% Retained	Cumulative Percentage Retained	% Passed
4.75	17.88	4.47	4.47	95.93
2	3.91	0.98	5.45	94.55
1.18	5.03	1.26	6.71	93.29
0.425	112.15	28.04	34.75	65.25
0.3	76.52	19.13	53.88	46.12
0.15	67.53	16.88	70.76	29.24
0.075	12.5	3.13	73.89	26.11
<0.075	104.48	26.12	100	0

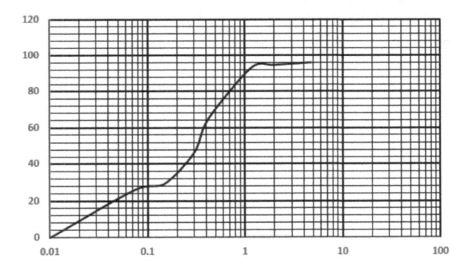

FIGURE 4.7 Graph showing D_{60}, D_{50}, and D_{10} values.

FROM THE GRAPH, $D_{60} = 0.39$; $D_{50} = 0.32$; $D_{10} = 0.023$

$C_u = D_{60}/D_{10} = 0.39/0.023 = 16.96$

Since $C_u \geq 2.0$, we should use P-A ratio for designing the gravel pack lying between 12 and 15.5

At P-A ratio = 12, D_{50} of gravel pack/D_{50} of aquifer material = 12

D_{50} of gravel pack/0.32 = 12; D_{50} of gravel pack = 0.32 × 12 = 3.84 mm

At P-A ratio = 15.5, D_{50} of gravel pack/D_{50} of aquifer material = 15.5

D_{50} of gravel pack/0.32 = 15.5; D_{50} of gravel pack = 0.32 × 15.5 = 4.96 mm

It implies that the D_{50} of the gravel pack should lie between the limiting values of 3.84 and 4.96 mm

To get the limiting curves of the grain size distribution of the gravel pack, mark a straight line from the vertical axis of the grain size distribution curve of the aquifer materials corresponding to the D_{50} of the material. Mark out the D_{50} values determined above (3.84 and 4.96 mm) and draw a curve parallel to the central portion of the grain size distribution curve of the aquifer material. Extend the curves to axis zero on the lower horizontal axis and up to 100% on the upper horizontal axis. Read off the sieve sizes corresponding to these optimum points. The gravel pack size should lie between these optimum points and should be screened to lie within that range. The screen size for the slots of the tube well should have sizes slightly higher than the size of material corresponding to the D_{10} of the gravel pack curve.

REFERENCES

Gravel Pack Design Studies. 1971. PAU, Ludhiana.

Gupta, S.N. 1964. Model experiments on steady filtering action in tube wells at the Irrigation Research Institute, Roorkee.

Johnson, Edward E. Inc., 1996. The principle and practical methods of developing water wells. *Technical Bulletin* 1033, Dec St. Paul. Minnesota, U.S.A., Johnson.

Johnson Division, 1966. Selection of screen slot size for uniform sand.

Smith, H.F., 1954. *Gravel Packing Water Wells*, Illinois Dept Registration and Education Circ., 44.

Part 3

Measurement of Aquifer Properties

5 Estimation of Specific Yield and Specific Retention

Objective: Estimation of Specific Yield and Specific Retention

5.1 PRINCIPLE

Groundwater fills the whole interstice spaces in the zone of saturation as a consequence the porosity coefficient brings forth an undeviating estimate of the water accommodated per unit volume. Drainage or well pumping can remove a portion of water from subterranean strata, but molecular and surface tension forces keep the water remnant in order.

5.2 DEFINITIONS OF RELATED TERMS

5.2.1 SPECIFIC RETENTION

It is defined as the ratio of volume of water of a soil or rock which it will retain after saturation against the force of gravity to its own volume. It is denoted by S_r. Thus,

$$S_r = \frac{W_r}{V} \tag{5.1}$$

where W_r is the volume occupied by retained water and V is the bulk volume of the soil or rock (Subramanya, 2013).

5.2.2 SPECIFIC YIELD

It is defined as the ratio of the volume of water of the soil or rock which after saturation can be drained by gravity to its own volume. It is denoted by S_y. Thus,

$$S_y = \frac{W_y}{V} \tag{5.2}$$

where W_y is the volume of water drained.

Values of S_r and S_y can be expressed as percentages because W_r and W_y constitute the total water volume in a saturated material. It is apparent that porosity (η) is given by:

$$\eta = W_r + W_y \tag{5.3}$$

DOI: 10.1201/9781003319757-8

TABLE 5.1
Range of Specific Yield of Common Earth Materials

Material	Range of Specific Yield (%)
Clay	1–18
Silt	1–40
Loess	14–22
Eolian sand	32–47
Sand (fine)	1–46
Sand (medium)	16–46
Sand (coarse)	18–43
Gravel (fine)	13–40
Gravel (medium)	17–44
Gravel (coarse)	13–25
Shale	0.5–5
Siltstone	1–33
Sandstone (fine-grained)	2–40
Sandstone (medium-grained)	12–41
Limestone and dolomite	0–36
Karstic limestone	2–15
Fresh granite	<0.1
Weathered granite	0.5–5
Fractured granite	2–10
Vesicular basalt	5–15

Formally, it is defined as "the volume of water that an unconfined aquifer releases from storage per unit surface area of aquifer per unit decline in the water table".

Specific yield can be measured by a well-pumping test method (Todd and Mays, 2005). Specific yield values of common earth materials are shown in Table 5.1.

5.3   LABORATORY PROCEDURE FOR DETERMINING SPECIFIC YIELD AND SPECIFIC RETENTION

a. Simple saturation and drainage:

In this method, the water yield is estimated by draining by gravity the columns of saturated material. Hence, the volume of water yielded is either determined directly or else evaluated by the porosity or moisture content obtained after draining by gravity.

b. Correlation with particle size

This method employs the estimation of either the effective size or the median diameter of the sample. After their computation, the values are compared with the curve which expresses the relation between effective size or median diameter and specific retention. Hence, the value of specific retention can be determined (Madhulika, 2018).

5.4   CALCULATIONS

1. Total volume of container soil + water at saturated condition (V) _____ Cu cm
2. Volume of water drained by gravity from this sample (W_y) _____ Cu cm
3. Volume of water retained in soil sample can be calculated by the weight basis (wet and dry weight of the sample) (W_r) _____ Cu cm

5.5   RESULTS

1. $S_r =$
2. $S_y =$
3. $\eta =$

Example 5.1

What will be the average drawdown over an area where 25 Mm³ of water has been pumped through various uniformly distributed wells having an area of 150 km² with specific yield of unconfined aquifer as 25%.

Solution:

Volume of water drained $= 25 \times 10^6$ m^3
Area $= 150$ km$^2 = 150 \times 10^6$ m^2
Specific yield $= 25\% = 0.25$
W.K.T.,

$$S_y = \frac{\text{Volume of water drained}}{\text{Volume of aquifer}}$$

$$0.25 = \frac{25 \times 10^6}{\text{Volume of aquifer}}$$

Volume of aquifer $= 10^8$ m^3

Volume of aquifer = Area $\times$ Depth

$$10^8 = 150 \times 10^6 \times \text{Depth}$$

Depth $= 0.66$ m

Example 5.2

The water table dropped by 4.5 m in an area of 100 ha. What will be the specific yield of aquifer if the porosity is 30% while specific retention is 10%. Also, estimate change in groundwater storage.

Solution:

Porosity = Specific Yield + Specific Retention

$\eta = S_y + S_r$

$0.3 = S_y + 0.1$

$S_y = 0.2 = 20\%$

Volume of water pumped out = Area $\times S_y \times$ Drop in water table

Volume of water pumped out = $100 \times 10^4 \times 0.2 \times 4.5$

Volume of water pumped out = $9 \times 10^5 \text{m}^3$

Example 5.3

The water table was initially at 25 m below ground level in a phreatic aquifer stretching over 1 km². The water table rose to a depth of 24 m below ground level eventually after irrigation with 20 cm depth of water. However, the water table dropped to 26.2 m below ground level afterwards when $3 \times 10^5 \text{m}^3$ of water was pumped out. Estimate specific yield of the aquifer and deficit in soil moisture (below field capacity) before irrigation.

Solution:

Area of aquifer = $1 \text{km}^2 = 10^6 \text{ m}^2$

Volume pumped out = $3 \times 10^5 \text{m}^2$

Drop in water table = 26.2 m – 24 m = 2.2 m

i. Volume of water pumped out = Area $\times$ Drop in Water Table $\times$ Specific Yield

$3 \times 10^5 = 10^6 \times 2.2 \times$ Specific Yield

Specific Yield = 0.136 = 13.6%

ii. Volume of irrigation water recharging the aquifer :

Area of Aquifer $\times$ Rise in Water Table $\times$ Specific Yield

Rise in Water table = 25 m – 24 m = 1 m

Considering an area of 1 m² of the aquifer:

Volume of irrigation water recharging the aquifer = $1 \times 1 \times 0.136$

Volume of irrigation water recharging the aquifer = 0.136 m = 136 mm

Soil moisture deficit below filed capacity before irrigation:

Depth of irrigation water – Volume of irrigation water recharging the aquifer

Thus, soil moisture deficit = 200 mm – 136 mm = 64 mm

REFERENCES

Madhulika. 2018. Specific Yield of an Aquifer Groundwater. https://www.geographynotes.com/aquifer/how-to-calculate-the-specific-yield-of-an-aquifer-groundwater/6810

Subramanya, K. 2013. "Groundwater". *Engineering Hydrology*, Shukti Mukherjee, Sandhya Chandresekhar and Sohini Mukherjee (Eds.), Fourth Edition, McGraw Hill Education Private Limited, New Delhi, India, 389–431.

Todd, K.D. and Mays, W.L. 2005. "Groundwater movement". *Groundwater Hydrology*, Bill Zobrist, Jennifer Welter and Valerie A. Vargas (Eds.), Third Edition, John Wiley and Sons Inc, Hoboken, NJ, 86–91.

6 Evaluation of Hydraulic Properties of Aquifer by Theis Method

Objective: Evaluation of Hydraulic Properties of Aquifer by Theis Method

6.1 UNSTEADY-STATE FLOW

When the flow conditions at any given time are not constant, the flow is said to be unsteady, that is,

$$\frac{dv}{dt} \neq 0 \qquad (6.1)$$

An example for unsteady flow is water flowing through a pipe of fluctuating diameter under changing pressure because of either expanding or diminishing water level of the reservoir or opening or conclusion of a valve or halting or beginning of hydraulic machines associated with the pipe.

Although the hydraulic conductivity and transmissivity of confined and unconfined aquifers can be easily determined using Thiem's steady-state equations, field conditions may be such that reaching steady-state flow takes a long time, necessitating the determination of aquifer properties under unsteady flow conditions (Subramanya, 2013).

6.2 UNSTEADY-STATE FLOW TO WELLS IN UNCONFINED AQUIFERS

Dewatering of the pore space in an unconfined aquifer with a dropping water table is not instantaneous but continues for some time following the decline in unsteady-state flow. Despite being unsaturated, the land above the water table continues to feed water to the declining water table. As a result, the particular yield grows at a decreasing rate as the pumping time increases. As a result, the saturation thickness of unconfined aquifers varies greatly. The aquifer parameters can be estimated using the approach used for unsteady-state flow in confined aquifers, assuming $s' = s - s^2/2H$, m, where s' is the drawdown component for the decrease in saturation thickness of the unconfined aquifer.

DOI: 10.1201/9781003319757-9

6.3 UNSTEADY-STATE FLOW TO WELLS IN CONFINED AQUIFERS

Theis (1935) introduced the time factor and storage coefficient to build a solution for determining aquifer parameters under unsteady-state flow conditions. When a well penetrates an extensive confined aquifer and is pumped at a steady rate, the discharge's influence increases outward with time. The discharge is equal to the rate of fall of the head multiplied by the storage coefficient aggregated over the area of influence. The head will continue to fall as long as the aquifer is functionally unlimited because the water must come from a reduction in storage inside the aquifer. As a result, there is an unstable flow. However, as the sphere of influence grows, the pace of decline slows down.

Theis equation for non-steady-state flow in aquifers, which is derived from the analogy between groundwater flow and heat conduction, is based on the following assumptions, which are in addition to the Thiem–Dupuit equations:

1. There is a confined aquifer.
2. The flow to a well is in an unstable state, meaning that neither the drawdown difference nor the hydraulic gradient is constant over time.
3. As the head drops, the water collected from storage is discharged instantly.
4. Because the well diameter is so small, the storage in the well can be ignored.

Jacob (1940) proposed the following differential equation governing the unsteady-state radial flow in a non-leaky confined aquifer in polar coordinates:

$$\frac{\partial^2 h}{\partial r^2} + \frac{l}{r}\frac{\partial h}{\partial r} = \frac{S}{T}\frac{\partial h}{\partial t} \tag{6.2}$$

where

T: Transmissibility of the aquifer, m²/s;
S: Storage coefficient, dimensionless;
R: Radial distance of the piezometer from the center of the pumped well, m;
t: Elapsed time after pumping is started, s.

Theis (1935) obtained the solution based on the analogy between groundwater flow and heat flow condition, and boundary conditions $h = h_0$ before pumping and $h \to h_0$ as $r \to \infty$ as pumping begins ($t \geq 0$) may be written as

$$h_0 - h = \frac{Q}{4\pi T}\int_{r^2 s/4tT}^{\infty} \frac{e^{-u}}{u}\,du \tag{6.3}$$

where $u = \dfrac{r^2 S}{4Tt}$

Q: Constant discharge rate, m³/s.

The exponential integral is written symbolically as $W(u)$, which in this usage is generally read as "well function of u" or "Theis' well function (Subramanya, 2013)".

Equation (6.3), in terms of the Theis well function, may be written as:

$$s = \frac{Q}{4\pi T} W(u) \qquad (6.4)$$

s: The unsteady-state drawdown, m.

6.4 PROCEDURE FOR DETERMINING HYDRAULIC PROPERTIES OF CONFINED AQUIFERS

The following are the step-by-step processes for determining the hydraulic characteristics of restricted aquifers:

1. Using the table of well functions of Theis, create a "type curve" (Figure 6.1) of the Theis well function by graphing values of $W(u)$ against u on the double logarithmic paper (1935).
2. On another double logarithmic paper, plot the values of s against t/r^2 on the same scale as the type curve. The type curve is overlaid with the observed data plot. The place of optimum match between the data plot and type curve (Figure 6.2) is found by keeping the coordinate axes of both data plot and type curve parallel.
3. On the overlapping region of the two sheets of graph papers, an arbitrary match point A is chosen, and the coordinates $W(u)$, $1/u$, s, and t/r^2 for this match point are computed. If the point is chosen when the type curve's coordinates are $W(u)=1$ and $1/u=10$, the calculation is substantially simplified.
4. The values of $W(u)$, s, and Q are substituted to give the value of transmissibility.

$$T = \frac{Q}{4\pi s} W(u) \qquad (6.5)$$

5. The value of S is calculated by substituting the values of T, t/r^2, and u in Eq. (6.4), that is,

$$S = 4Tu\frac{1}{r^2}$$

Example 6.1

Calculate the hydraulic properties of an aquifer using Theis method. The pumping test data are given in Table 6.1.
   And the value of $1/u = 10$. The value of drawdown s on $A = 0.15$ m and t/r^2 on $A = 1.5 \times 10^{-3}$ min/m². Introduction of these values in Eqs. (7.3) and (7.4) gives:

$$T = \frac{0.006}{4 \times 3.14 \times 0.15} \times 1 = 0.00318 \ \text{m}^2/\text{min}$$

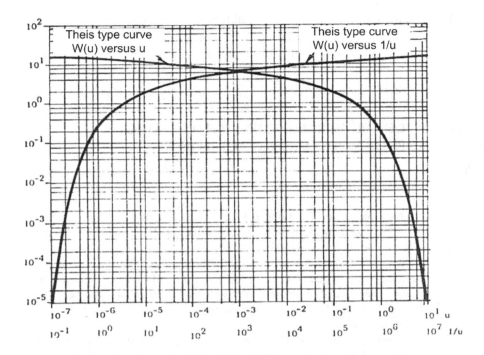

FIGURE 6.1   Theis-type curve for $W(u)$ versus u and $W(u)$ versus $(1/u)$.

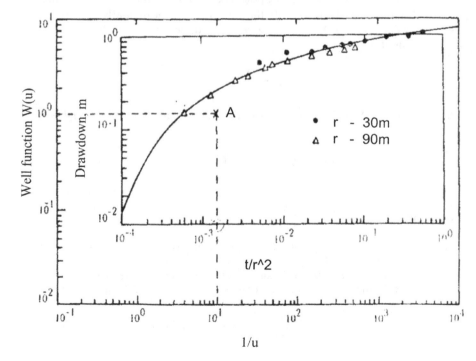

FIGURE 6.2   Plot of s versus t/r^2 superimposed on Theis-type curve.

TABLE 6.1
Pumping Test Data with Constant Rate of Discharge $Q = 0.006\,m^3/min$

Elapsed Time t, (min)	Drawdown, s (m)		T/r^2	
	Distance of Observation Well from Pumped Well, r			
	30 m	90 m	30 m	90 m
0	0	0	0	0
5	0.490	0.130	0.005	0.0006
10	0.680	0.207	0.011	0.0012
20	0.700	0.306	0.022	0.0024
30	0.750	0.365	0.033	0.0036
50	0.795	0.427	0.055	0.0060
60	0.820	0.450	0.066	0.0070
90	0.870	0.495	0.100	0.0110
180	0.934	0.570	0.200	0.0220
300	0.990	0.613	0.333	0.0370
480	1.050	0.700	0.533	0.0590
600	1.053	0.704	0.666	0.0740

and

$$S = 4 \times 0.00318 \times \frac{1}{10} \times 1.5 \times 10^{-3} = 1.91 \times 10^{-6}$$

6.5   AQTESOLV

AQTESOLV is the world's driving software for the plan and interpretation of aquifer tests in any type of aquifer such as confined, unconfined, leaky, or fractured.

Started around in 1989, AQTESOLV has been inseparable from greatness in the understanding of aquifer test information.

Ensuing work by Hantush (1961a, b) stretched out the Theis method to address to some extent partially penetrating wells. Whenever we select the Theis arrangement in AQTESOLV, we might investigate information from both fully and to some degree partially penetrating wells.

AQTESOLV consolidates the rule of superposition so as to incorporate variable rate and recuperation tests with this technique.

6.6   DATA REQUIREMENTS

AQTESOLV requires the following data:

- The location of pumping and observation well
- The rate of pumping
- The time and displacement of well

- The depth of partial penetration
- Saturation thickness
- Anisotropy ratio for partially penetrating wells

6.7   ESTIMATED PARAMETERS

AQTESOLV gives visual and programmed strategies to match the Theis non-equilibrium method to get information from pumping tests and recuperation tests. The assessed aquifer properties are as follows:

- Transmissivity
- Storativity
- Hydraulic conductivity
- Saturated aquifer thickness

REFERENCES

AQTESOLV.1989. http://www.aqtesolv.com/theis.htm

Hantush, M.S., 1961a. Drawdown around a partially penetrating well, *J. Hydraul. Div.*, vol. 87, no. 4, pp. 83–98.

Hantush, M.S., 1961b. Aquifer tests on partially penetrating wells, *J. Hydraul. Div.*, vol. 87, no. 5, pp. 171–194.

Jacob, C.E., 1940. On the flow of water in an elastic artesian aquifer. *Transactions American Geophysical Union*, vol 21, pp. 574–586.

Subramanya, K. 2013 "Groundwater". *Engineering Hydrology*, Shukti Mukherjee, Sandhya Chandresekhar and Sohini Mukherjee (Eds.), Fourth Edition, McGraw Hill Education Private Limited, New Delhi, India.

Theis, C.V., 1935. The relation between the lowering of the piezometric surface and the rate and duration of discharge of a well using groundwater storage, *Am. Geophys. Union Trans.*, vol. 16, pp. 519–524.

7 Evaluation of Hydraulic Properties of Aquifer by the Cooper– Jacob Method

Objective: Evaluation of Hydraulic Properties of Aquifer by Cooper–Jacob Method

7.1 COOPER–JACOB METHOD OF SOLUTION

Hilton Hammond Cooper (1913–1990) and Charles Edward Jacob (1914–1970) worked as groundwater hydrologists with the US Geographical Survey and formulated a broadly involved graphical method for calculating hydraulic properties (transmissivity and storativity) of non-leaky aquifers.

The Cooper and Jacob (1946) arrangement (occasionally called as Jacob's adjusted non-equilibrium technique) is a late-time estimation acquired from Theis type curve method. Investigation with the Cooper and Jacob strategy includes matching a straight line to drawdown information plotted as a component of the logarithm of time since pumping started.

Cooper and Jacob (1946) discovered that when r is small and t is large, u is so small that the series of $W(u)$ becomes negligible after the first two terms. As a result, the following relationship can be used to approximate the drawdown:

$$s = \frac{Q}{4\pi T}(-0.5772 - \log_e \frac{r^2 S}{4Tt})$$

$$s = \frac{Q}{4\pi T}(\log_e \frac{r^2 S}{4Tt}) - 0.5772) \tag{7.1}$$

This reduces to:

$$s = \frac{2.30}{4\pi} \frac{Q}{T} \log_{10} \frac{2.25}{r^2} \frac{Tt}{r^2 s} \tag{7.2}$$

If s_1 and s_2 are the drawdowns at time t_1 and t_2, since pumping started

$$s_1 - s_2 = \frac{2.30}{4\pi} \frac{Q}{T} \log_{10}(\frac{t_2}{t_1}) \tag{7.3}$$

DOI: 10.1201/9781003319757-10

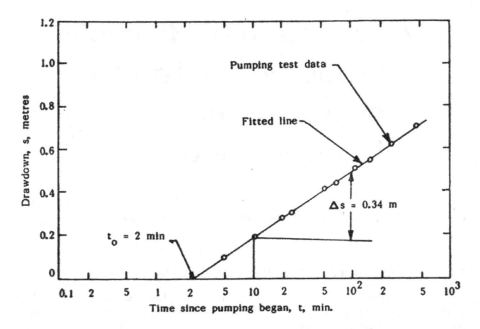

FIGURE 7.1 Cooper–Jacob method for solution of non-equilibrium equation.

If time drawdown data on a pumping well is plotted on a semi-log paper (Figure 7.1) and for convenience t_1 and t_2 are chosen one log cycles apart, then:

$$\log_{10} \frac{t_2}{t_1} = 1, \text{ and if } s_1 - s_2 = \Delta s, \text{ then}$$

$$\Delta s = \frac{2.30\, Q}{4\pi\, T} \tag{7.4}$$

or

$$T = \frac{2.30\, Q}{4\pi\, \Delta s} \tag{7.5}$$

We know that,

$$S = 0 \text{ when } \log_{10} \frac{2.25\, T t}{r^2\, s} = 0$$

that is, when $\dfrac{2.25\, T t}{r^2\, S} = 1$.

Therefore, a plot of drawdown s versus the logarithm of t forms a straight line. Projecting this line to $s=0$, where $t=t_0$, the times for $s=0$ can be noted and S can be computed a (Subramanya, 2013):

$$s = \frac{2.25\, T\, t_0}{r^2} \qquad (7.6)$$

Example 7.1

Calculate the values of transmissivity and coefficient of storage using the pumping test data of Table 7.1 for $Q=0.006$ m³/min and $r=90$ m, adopting the Cooper–Jacob method.

Solution:

The pumping test data of Table 7.1, s and t for $r=90$ m are plotted on semi-log paper (Figure 7.1). From the straight line fitted through the points, it is observed that

$$S = 0.34 \text{ m and } t_0 = 2 \text{ min.}$$

From Eq. (7.5),

$$T = \frac{2.30\, Q}{4\pi\, \Delta s}$$

$$Q = 0.006 \text{ m}^3/\text{min} = 8.64 \text{ m}^3/\text{day}$$

$$T = \frac{2.30 \times 8.64}{4\pi\,(0.34)} = 4.65 \text{ m}^2 / \text{day}$$

TABLE 7.1
Pumping Test Data with Constant Rate of Discharge $Q=0.006$ m³/min

| | Drawdown, s (m) | | T/r^2 | |
| | Distance of Observation Well from Pumped Well, r | | | |
Elapsed Time, t (min)	30 m	90 m	30 m	90 m
0	0	0	0	0
5	0.490	0.130	0.005	0.0006
10	0.680	0.207	0.011	0.0012
20	0.700	0.306	0.022	0.0024
30	0.750	0.365	0.033	0.0036
50	0.795	0.427	0.055	0.0060
60	0.820	0.450	0.066	0.0070
90	0.870	0.495	0.100	0.0110
180	0.934	0.570	0.200	0.0220
300	0.990	0.613	0.333	0.0370
480	1.050	0.700	0.533	0.0590
600	1.053	0.704	0.666	0.0740

and from Eq. (7.6)

$$S = \frac{2.25\, T\, t_0}{r^2}$$

$$S = \frac{2.25 \times 4.65 \times 2}{60 \times 24 \times 90 \times 90}$$

$$S = 1 \times 10^{-6}$$

Example 7.2

A 30 cm well penetrating a confined aquifer is pumped at a rate of 1,200 lpm. The drawdown at an observation well at a radial distance of 30 m is as follows:

Time from Start (min)	Drawdown (m)
1	0.2
2.5	0.5
5	0.8
10	1.2
20	1.8
50	2.5
100	3.0
200	3.7
500	4.4
1,000	5.0

The drawdown is plotted against time on a semi-log plot (Figure 7.2). It is seen for $t > 10$ minutes, the drawdown values describe a straight line. A best-fitting straight line is drawn for data points with $t > 10$ minutes. From this line:
When $S = 0$, $t = t_0 = 2.5$ minutes $= 150$ seconds
$S_1 = 3.1$ m at $t_1 = 100$ minutes
$S_2 = 5.0$ m at $t_2 = 1,000$ minutes
Also, $Q = 1,200$ lpm $= 0.02$ m/s
W.K.T.

$$s_2 - s_1 = \frac{Q}{4\pi T} \ln \frac{t_2}{t_1}$$

$$5 - 3.1 = \frac{0.02}{4\pi T} \ln \frac{1,000}{100}$$

$$T = \frac{0.02}{4 \times \pi \times 1.9} \ln(10)$$

$$T = 1.929 \times 10^{-3}\, \text{m}^3/\text{s/m} = 1.67 \times 10^{-5}\, \text{lpd/m}$$

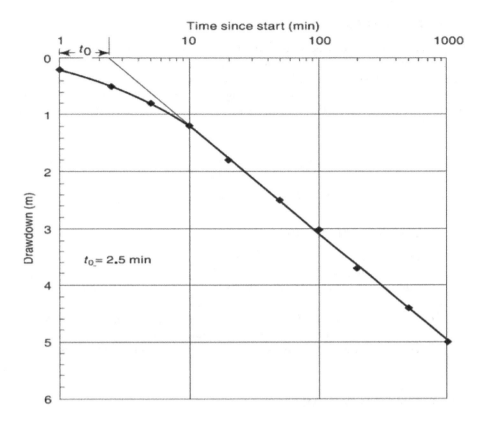

FIGURE 7.2 Drawdown plotted against time.

Also,

$$S = \frac{2.25\ Tt_0}{r^2}$$

$$S = \frac{2.25 \times 1.929 \times 10^{-3} \times 150}{30^2}$$

$$S = 7.23 \times 10^{-4}$$

7.2 AQTESOLV

AQTESOLV as described in the previous chapter can be used to analyze the hydraulic properties of aquifer using the Cooper and Jacob method.

The variable-rate execution of the Cooper and Jacob arrangement in AQTESOLV compares to the method of Birsoy and Summers (1980).

7.2.1   Radius of Influence

The radius of influence is defined as the outspread distance from a pumping well where drawdown is actually zero for a given time frame.

Bear (1979) shows that the Cooper and Jacob condition for drawdown prompts the accompanying basic equation for figuring the radius of influence, $R(t)$, of a completely penetrating well releasing at a steady rate in an endless non-leaky confined aquifer:

$$R_t = 1.5 \sqrt{\frac{Tt}{S}}$$

Radius of influence can be calculated with ease using AQTESOLV.

7.3   DATA REQUIREMENTS

- Location of pumping and observation wells
- The rate of pumping
- The measurements of observation wells

The Cooper and Jacob arrangement is confined to one pumping well. On the off chance that AQTESOLV informational index contains numerous pumping wells, the Cooper and Jacob arrangement just uses the first pumping well for the investigation of analysis data.

7.4   ESTIMATED PARAMETERS

AQTESOLV gives visual and programmed strategies to match the Cooper and Jacob altered non-equilibrium technique to pump test information. The assessed aquifer properties are as per the following:

- Transmissivity
- Storativity

REFERENCES

AQTESOLV.1989. http://www.aqtesolv.com/theis.htm

Bear, J. 1979. "Hydraulics of groundwater". *McGraw-Hill Series in Water Resources and Environmental Engineering*. McGraw-Hill, New York.

Birsoy, Y.K. and Summers, W.K. 1980. Determination of aquifer parameters from step tests and intermittent pumping, *Ground Water*, vol. 18, no. 2, pp. 137–146.

Cooper, H.H. and Jacob, C.E. 1946. A generalized graphical method for evaluating formation constants and summarizing well field history, *Am. Geophys. Union Trans.*, vol. 27, pp. 526–534.

Subramanya, K. 2013. "Groundwater". *Engineering Hydrology*. Shukti Mukherjee, Sandhya Chandresekhar and Sohini Mukherjee (Eds.), Fourth Edition, McGraw Hill Education Private Limited, New Delhi, India.

8 Evaluation of Hydraulic Properties of Aquifer under Unsteady-State Condition by the Chow Method

Objective: Evaluation of Hydraulic Properties of Aquifer by the Chow Method under Unsteady-State Condition

8.1 THEORY

Chow (1952) derived the equation for determining the aquifer's properties on the basis of Theis non-equilibrium equation. He introduced a different type of function, that is, $F(u)$, which is a function of u. The values of $W(u)$ and u corresponding to known drawdowns at any specified time (t) can be obtained against $F(u)$. The nomograph showing the relationship among $F(u)$, u, and $W(u)$ is shown in Figure 8.1.

The Chow technique has the benefit of not requiring curve fitting and having no limitations in terms of application. The Cooper–Jacob approach has a similar requirement for pump test data and visualization. A random point is picked on the plotted curve (Figure 8.1) and the coordinates of t and s are noted. The drawdown difference in meters per log cycle of the time is calculated by drawing a tangent to the curve at the specified point. The value of $F(u)$ from $F(u) = \dfrac{s}{\Delta s}$ and then the corresponding values of $W(u)$ and u are computed using Figure 8.1. Finally the formation constants T and s are computed (Subramanya, 2013).

Example 8.1

Determine the formation constants using Chow's method, from the pump test data given in Table 8.1, for $r = 90$ m and $Q = 0.006$ m³/min (8.64 m³/day).

Solution:

The pumping test data given in Table 8.1 are plotted on semi-log paper (Figure 8.2).

DOI: 10.1201/9781003319757-11

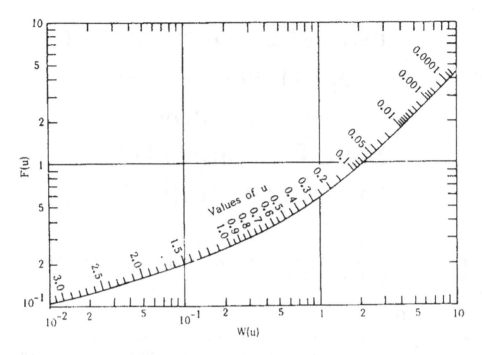

FIGURE 8.1   Relationship between $F(u)$, $W(u)$, and u.

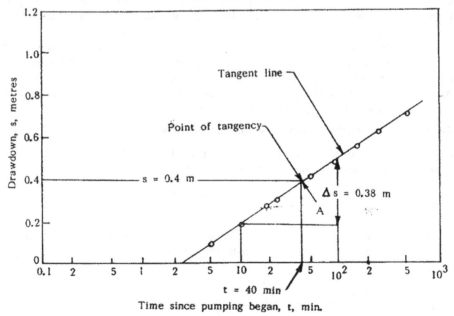

FIGURE 8.2   Chow's method for solution of the non-equilibrium equation for flow in confined aquifer.

TABLE 8.1
Pumping Test Data

Elapsed Time, t (min)	Drawdown, s (m)		T/r^2	
	Distance of Observation Well from Pumped Well, r			
	30 m	90 m	30 m	90 m
0	0	0	0	0
5	0.490	0.130	0.005	0.0006
10	0.680	0.207	0.011	0.0012
20	0.700	0.306	0.022	0.0024
30	0.750	0.365	0.033	0.0036
50	0.795	0.427	0.055	0.0060
60	0.820	0.450	0.066	0.0070
90	0.870	0.495	0.100	0.0110
180	0.934	0.570	0.200	0.0220
300	0.990	0.613	0.333	0.0370
480	1.050	0.700	0.533	0.0590
600	1.053	0.704	0.666	0.0740

Point A is selected on the curve arbitrarily where $t=40$ minutes and $s=0.40$. The drawdown difference per log cycle of time is,
$\Delta s = 0.38$ m, then $F(u) = 0.40/0.38 = 1.05$ and from Figure 8.2: $W(u) = 2.20$, $u = 0.05$

$$T = \frac{Q}{4\pi\Delta s}W(u) = \frac{8.64}{4\pi \times 0.38} \times 2.2 = 3.98 \text{ m}^3 / \text{day}$$

$$S = \frac{4Ttu}{r^2} = \frac{4 \times 3.98 \times 40 \times 0.05}{24 \times 60 \times 90 \times 90} = 2.72 \times 10^{-6}$$

REFERENCES

Subramanya, K. 2013 "Groundwater". *Engineering Hydrology*, Shukti Mukherjee, Sandhya Chandresekhar and Sohini Mukherjee (Eds), Fourth Edition, McGraw Hill Education Private Limited, New Delhi, India.

Ten Chow, V. 1952. On the determination of transmissibility and storage coefficients from pumping test data, *Eos Trans. Am. Geophys. Union*, vol. 33, pp. 397–404.

9 Evaluation of Hydraulic Properties of Aquifer by Recovery Test

Objective: Evaluation of Hydraulic Properties of Aquifer by Recovery Test

9.1 RECOVERY TEST

The water level in the well and observation wells begins to rise when the pump is turned off at the end of a pumping test. This is referred to as groundwater level recovery. Residual drawdown is the drop in water level (drawdown) below the original static water level (before pumping) and during the recovery period. A schematic illustration of the change in water level over time during and after pumping is shown

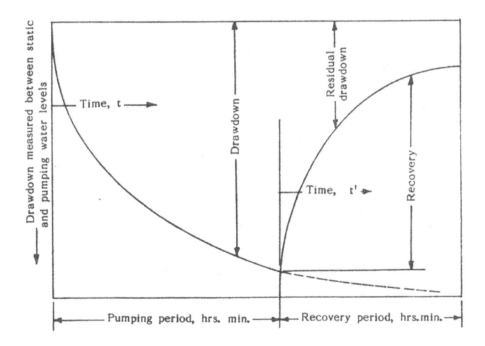

FIGURE 9.1 Definition sketch illustrating the drawdown and recovery curves in an observation well near a pumping well.

DOI: 10.1201/9781003319757-12

in Figure 9.1. By assessing the residual drawdown, the aquifer's transmissibility can be determined, providing an independent check on the pumping test results.

The residual drawdown can be calculated as follows (Theis, 1935):

$$s' = \frac{Q}{4\pi T}(W(u) - W(u')) \tag{9.1}$$

where

$$u = \frac{r^2 S}{4Tt} \text{ and } u' = \frac{r^2 S}{4Tt'} \tag{9.2}$$

Figure 9.1 defines t and t'. For small values of r and large values of t', Eq. (9.1) can be approximated as

$$s' = \frac{2.30}{4\pi}\frac{Q}{T}\log\frac{t}{t'} \tag{9.3}$$

The residual drawdown s' versus $\frac{t}{t'}$ are plotted on a semi-logarithmic paper. The slope of the straight line so plotted equals 2.30 $Q/4\pi T$, so that for $\Delta s'$, the residual drawdown per log cycle of t/t', the transmissibility becomes

$$T = \frac{2.30\,Q}{4\pi\Delta s} \tag{9.4}$$

The recovery test method cannot be used to determine the comparable values of s (Subramanya, 2013).

Example 9.1

Calculate the value of transmissivity using the recovery test data given in Table 9.1, where the uniform rate of pumping may be assumed as 2,000 m³/day. Pumping was shut down after 200 minutes. Thereafter measurements of s' and t' were made as tabulated in Table 9.1.

The recovery test method for solution of the non-equilibrium equation is given in Figure 9.2.

Solution:

The values of t/t' are computed as shown in Table 9.1 and then plotted versus s' on semi-log paper (Figure 9.2). A straight line is fitted through the points.

The values of $\Delta s' = 0.38$ m

$$T = \frac{2.30\,Q}{4\pi\Delta s'} = \frac{2.30\times(2,000)}{4\pi\times(0.38)} = 963.8 \text{ m}^3/\text{day}$$

TABLE 9.1

Recovery Test Data to Determine Transmissivity of Aquifer

t' (min)	t (min)	t/t'	S' (m)
1	200	200	0.80
2	202	101	0.68
3	203	68	0.60
5	205	41	0.50
7	207	29	0.45
10	210	21	0.41
15	215	14	0.39
2	220	13	0.35
30	230	8	0.30
40	240	6	0.25
60	260	4.5	0.22
80	280	3.5	0.14
100	300	3	0.14
140	340	2.4	0.13
180	380	2.1	0.10

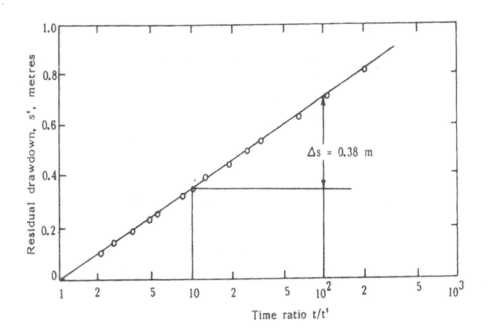

FIGURE 9.2   Recovery test method for solution of the non-equilibrium equation.

Example 9.2

Recovery test on a well in a confined aquifer yielded the following data: Pumping was at a uniform rate of 1,200 m²/day and was stopped after 210 minutes of pumping. Recovery data were as shown below:

Time Since Stoppage of Pump (min)	Residual Drawdown (m)
2	0.70
5	0.55
10	0.45
20	0.30
40	0.25
90	0.19
150	0.15
210	0.10

Estimate the transmissibility of the aquifer.

Solution:

Here, since,
$t_1 = 210$ minutes,
$t = t_1 + t' = 210 + t'$
The time ratio t/t' is calculated (as shown in the table below) and a semi-log plot of s' versus t/t' is plotted (Figure 9.3).

t'	t	t/t'	s'
2	212	106	0.70
5	215	43	0.55
10	220	22	0.45
20	240	11.5	0.30
40	250	6.25	0.25
90	300	3.33	0.19
150	360	2.4	0.15
210	420	2	0.10

A best-fitting straight line through the plotted points is given by the equation:

$$s' = 0.1461\ln(t/t') - 0.002$$

Therefore, slope of the best-fit line = 0.1461

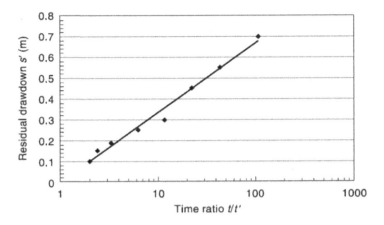

FIGURE 9.3   Plot of residual drawdown against time ratio (t/t').

$$\frac{Q}{4\pi T} = 0.1461$$

$$T = \frac{1,200}{0.1461 \times 4 \times \pi}$$

$$T = 654 \text{ m}^2/\text{day}$$

REFERENCES

Subramanya, K. 2013. "Groundwater". *Engineering Hydrology*, Shukti Mukherjee, Sandhya Chandresekhar and Sohini Mukherjee (Eds.), Fourth Edition, McGraw Hill Education Private Limited, New Delhi, India.

Theis, C.V., 1935. The relation between the lowering of the piezometric surface and the rate and duration of discharge of a well using groundwater storage, *Am. Geophys. Union Trans.*, vol. 16, pp. 519–524.

10 Study of Leaky and Non-Leaky Aquifers

Objective: Leaky Aquifers and Non-Leaky Aquifers

10.1 LEAKY AQUIFERS

Leaky aquifers are defined as aquifers that are overlain or underlain by semi-permeable layers. Vertical leakage or seepage through semi-confining formations into the aquifer can provide a major amount of the yield in such aquifers, as shown in Figure 10.1.

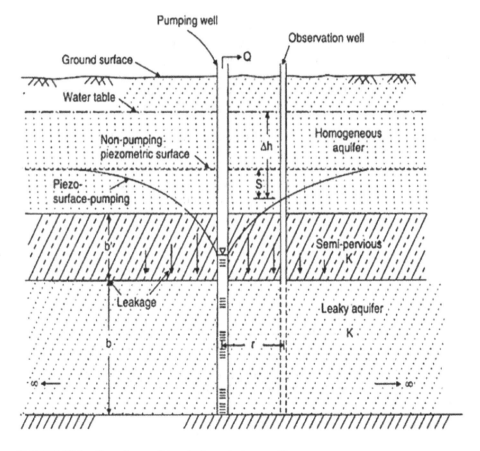

FIGURE 10.1 Pumping well in a leaky artesian aquifer.

DOI: 10.1201/9781003319757-13

Darcy's law states that the downward vertical flow velocity (v) through the semi-constricting layer (leakage rate, i.e., vertical discharge per unit area) is proportional to the difference between the groundwater table and the piezometric head in the confined aquifer (Subramanya, 2013).

$$V = \frac{K'}{\left(\dfrac{\Delta h}{b'}\right)} \tag{10.1}$$

And the total vertical leakage:

$$Q_c = v.A_c$$

$$Q_c = \frac{\dfrac{K'}{b'}}{\Delta h A_c} \tag{10.2}$$

where

Q_c = Leakage through the confining bed;
K' = Coefficient of vertical permeability of confining bed;
b' = Thickness of confining bed through which leakage occurs;
A_c = Area of confining bed through which leakage occurs; and
Δh = Difference between the piezometric head in the aquifer and the source bed above the confining bed (GWT in Figure 10.1).

The leakage (recharge) is proportional to the difference in head Δh and is more near the pumping well since the piezometric surface is very low. The ratio K'/b' is called leakance, and its reciprocal b'/K' is called the hydraulic resistance "c" of the confining layer and has the dimension of time; $\sqrt{TC}$ is called the leakage factor "B" and has the dimension of length; and T is the transmissibility ($= Kb$) of the confined aquifer.

Example 10.1

From the pumping tests of a semi-confined aquifer of thickness 30 m and permeability 20 m/day, it is estimated that the recharge rate from an overlying unconfined aquifer through an aquitard of thickness 2 m is 50 mm/year. The average piezometric surface in the semi-confined aquifer is 16 m below the water table in the unconfined aquifer. Determine the hydraulic characteristics of the aquitard (semi-confining layer) and the aquifer.

If a well drilled in the aquifer is pumped at the rate of 5,000 m³/day, how many square kilometers of recharge area are required to sustain the flow at the estimated recharge of 50 mm/year?

Solution:

Recharge through aquitard per unit area (1 m²)

$$Q = K'iA$$

$$\frac{0.050}{365} = K' \frac{16}{2} (1 \times 1)$$

$$K' = 1.71 \times 10^{-5} \text{m / day}$$

$$\text{Leakance} = \frac{K'}{b'} = \frac{1.71 \times 10^{-5}}{2} = 8.55 \times 10^{-6} \text{ /day}$$

$$\text{Hydraulic Resistance} = c = \frac{b'}{K'} = 1.17 \times 10^{-5} \text{days}$$

$$\text{Aquifer} \quad T = Kb = 20 \times 30 = 600 \text{ m}^2/\text{day}$$

$$\text{Leakage Factor} \quad B = \sqrt{TC} = \sqrt{600 \times 1.17 \times 10^5} = 8,380 \text{ m}$$

To sustain pumpage of $5,000$ m^3 / day , the recharge area A in km^2 is required

$$\frac{0.050}{365} A = 5,000$$

$$A = 36.5 \text{ km}^2$$

10.2 NON-LEAKY AQUIFERS

Recognizing the physical parallel between heat flow in solids and groundwater flow in porous media, Charles Vernon Theis (1900–1987) was the first hydrologist to build a rigorous mathematical model of transient flow of water to a pumping well.

The Theis (1935) solution (or Theis non-equilibrium approach) was a game-changing method for estimating the hydraulic parameters (transmissivity and storativity) of non-leaky confined aquifers. By matching the Theis-type curve to drawdown data shown as a function of time on double logarithmic axes, the Theis technique can be used to do the analysis.

Hantush (1961a) went on to extend Theis technique to account for partially penetrating wells

(AQTESOLV, 1989).

10.2.1 ASSUMPTIONS

- Aquifer has an infinite areal extent
- Aquifer is homogeneous and of uniform thickness
- Control well is fully or partially penetrating
- Flow to the control well is horizontal when the control well is fully penetrating
- Aquifer is non-leaky confined
- Flow is unsteady

- Water is released instantaneously from storage with decline of hydraulic head
- Diameter of a pumping well is very small so that storage in the well can be neglected

10.2.2 EQUATIONS

The Theis equation for flow to a fully penetrating line sink discharging at a constant rate in a homogeneous, isotropic, and non-leaky confined aquifer of infinite extent is as follows:

$$s = \frac{Q}{4\pi T} \int_u^\infty \frac{e^{-y}}{y} \, dy$$

$$u = \frac{r^2 S}{4Tt}$$

Groundwater hydrologists commonly refer to the above integral as the *Theis well function*, abbreviated as $w(u)$. Therefore, we may write the Theis equation in compact notation as follows:

$$s = \frac{Q}{4\pi T} w(u)$$

$$w(u) = -0.5772 - \ln(u) + u - \frac{u^2}{2.2!} + \frac{u^3}{3.3!} - \frac{u^4}{4.4!} + \ldots$$

A partially penetrating pumping well produces vertical components of flow in the aquifer. Hantush (1961b) derived equations extending the Theis method to include partial penetration effects in non-leaky confined aquifers. In the case of a piezometer, the following equation applies:

$$s = \frac{Q}{4\pi T} w(u) + \frac{2b^2}{\pi(l-d)} \sum_{n=1}^\infty \frac{1}{n} \left(\sin\left(\frac{n\pi l}{b}\right) - \sin\left(\frac{n\pi d}{b}\right) \right) \cdot \cos\left(\frac{n\pi z}{b}\right)$$

$$\beta = \sqrt{\frac{Kz}{Kr}} \frac{n\pi r}{b}$$

The following equation computes drawdown for a partially penetrating observation well:

$$s = \frac{Q}{4\pi T} w(u) + \frac{2b^2}{\pi^2(l-d)(l'-d)} \sum_{n=1}^\infty \frac{1}{n} \left(\sin\left(\frac{n\pi l}{b}\right) - \sin\left(\frac{n\pi d}{b}\right) \right) \cdot \left(\sin\left(\frac{n\pi l'}{b}\right) \right.$$

$$\left. - \sin\left(\frac{n\pi d'}{b}\right) \right) \cdot w(u, \beta)$$

where
- b is aquifer thickness [L]
- d is the depth to the top of pumping well screen [L]
- d' is the depth to the top of observation well screen [L]
- Kr is the radial (horizontal) hydraulic conductivity [L/T]
- Kz is the vertical hydraulic conductivity [L/T]
- l is the depth to the bottom of pumping well screen [L]
- l' is the depth to the bottom of observation well screen [L]
- Q is pumping rate [L^3/T]
- r is radial distance from pumping well to observation well [L]
- s is drawdown [L]
- S is storativity [dimensionless]
- t is elapsed time since start of pumping [T]
- T is transmissivity [L^2/T]
- $w(u)$ is the Theis well function for non-leaky confined aquifers [dimensionless]
- $w(u, \beta)$ is the Hantush and Jacob well function for leaky confined aquifers [dimensionless]
- y is a variable of integration
- z is piezometer depth [L]

At large distances, the effect of partial penetration becomes negligible when

$$r > \frac{1.5b}{\sqrt{\dfrac{Kz}{Kr}}}$$

10.3   DATA REQUIREMENTS

- Pumping and observation well locations
- Pumping rate(s)
- Observation well measurements (time and displacement)
- Partial penetration depths (optional)
- Saturated thickness (for partially penetrating wells)
- Hydraulic conductivity anisotropy ratio (for partially penetrating wells)

10.4   SOLUTION OPTIONS

- Variable pumping rates
- Multiple pumping wells
- Multiple observation wells
- Partially penetrating pumping and observation wells
- Boundaries

REFERENCES

AQTESOLV. 1989. http://www.aqtesolv.com/theis.htm

Hantush, M.S. 1961a. Drawdown around a partially penetrating well, *J. Hydraul. Div.*, vol. 87, no. 4, pp. 83–98.

Hantush, M.S. 1961b. Aquifer tests on partially penetrating wells, *J. Hydraul. Div.*, vol. 87, no. 5, pp. 171–194.

Subramanya, K. 2013. "Groundwater". *Engineering Hydrology*, Shukti Mukherjee, Sandhya Chandresekhar and Sohini Mukherjee (Eds.), Fourth Edition, McGraw Hill Education Private Limited, New Delhi, India.

Theis, C.V. 1935. The relation between the lowering of the piezometric surface and the rate and duration of discharge of a well using groundwater storage, *Am. Geophys. Union Trans.*, vol. 16, pp. 519–524.

Part 4

Wells and Their Design

11 Testing of Well Screen

Objective: Testing of Well Screen

11.1 THEORY

Strainer wells and slotted-pipe gravel-packed wells are two types of tube wells that use screens to allow water from the surrounding aquifer to enter. The borehole is drilled, and pipes and screens are dropped into it. The screens are positioned in the opposite direction of the water-bearing strata. At the bottom of the pipe and strainer, assembly is a bail plug. Percussion rigs are commonly used to drill strainer wells. When the casing is removed, the annular gap between the tube well assembly and the bore walls is filled with natural formation material. The well screen is made to fit the type of aquifer material found in the area. These wells are usually only drilled to a few feet deep. The slotted-pipe gravel-packed well is a variation on the strainer well in which the slotted-pipe screen is shrouded (Figure 11.1).

A variety of well screens are utilized depending on the aquifer's needs and the farmer's financial situation. The following are desirable elements for a properly designed well screen, according to Johnson (1966):

i. Continuous and uninterrupted openings in the shape of slots around the circumference of the screen.
ii. Keep slot openings close together to maximize the amount of free space.
iii. Inward-widening V-shaped slot apertures.
iv. Galvanic corrosion is avoided by using a single metal structure.
v. The ability to adapt to a variety of groundwater and aquifer conditions through the use of varied materials.
vi. Maximum open area while maintaining sufficient strength.
vii. Sufficient strength to withstand the force that may be applied to the screen during and after installation.
viii. A complete set of accessories to help with screen installation and well finishing.

11.2 STRAINER-TYPE SCREENS

Water-bearing strata with coarse sand and gravel particles are best served by strainer-type filters. Alluvial plains, coastal alluvium, and other places next to river banks include such formations. The water supply from the aquifer is obtained through a screen with narrow or extremely small apertures in this form of tube well. Sand particles that are larger than the entrance will not fit into the tube well. The strainers are placed against one or more layers that contain water. Continuous-slot strainers,

DOI: 10.1201/9781003319757-15

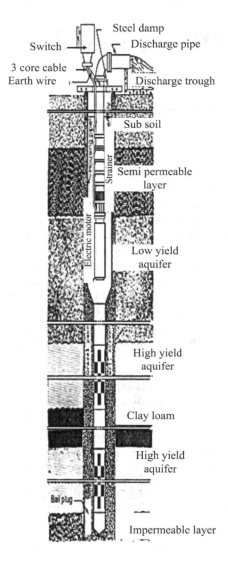

FIGURE 11.1 Arrangement of tube well assembly, gravel pack, and a submersible pump.

louver strainers, rope-wound strainers, and pipe-base strainers with fine-mesh net envelopes are all examples of strainer-type well screens (Boman *et al.*, 2006).

11.3 CONTINUOUS-SLOT SCREENS

Wire and rope are looped around a suitable frame constructed of rods or bars of iron or other materials in continuous-slot screens. It's typically created by spirally winding cold-drawn wire with a roughly triangular cross-section around a circular array of longitudinal rods (Figure 11.2). The wire is securely joined to the rod at

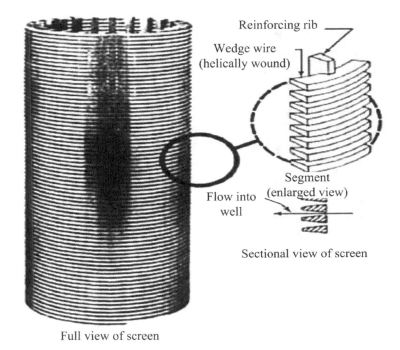

Reinforcing rib

Wedge wire
(helically wound)

Segment
(enlarged view)

Flow into
well

Sectional view of screen

Full view of screen

FIGURE 11.2 Continuous-slot wedge wire well screen.

each location where it crosses it. When compared to the slotted-pipe well screen, the well loss is significantly reduced. The inwardly widening slots prevent particles from striking inside the slot and obstructing it. As a result, the slots aren't clogged.

11.4 LOUVER OR SHUTTER-TYPE SCREENS

Louver or shutter-style well screens have rows of louvers as apertures. The apertures can be aligned at right angles or parallel to the screen's axis. Stamping is used to create them in the wall of a welded tube. Despite being more hydraulically efficient, continuous-slot and louver-type screens have not gained popularity in developing nations due to their high cost. The majority of strainer wells in India are shallow tube wells drilled by percussion (hand boring or mechanically operated cable-tool method). Rope or strands of nylon fiber are twisted around steel or bamboo frames to make the strainers. Agricultural strainers, made of brass wire netting wrapped around perforated tubing, are also utilized, though they aren't as widespread as they formerly were. In shallow tube wells, polyvinyl chloride (PVC) strainers made of nylon mesh net wrapped around perforated rigid PVC pipe are highly common.

11.5 COIR-ROPE STRAINER

A low-cost screen for shallow tube wells is the coir-rope strainer (Figure 11.3). It's made by spirally winding a 3–5 mm coir chord around a circular array of mild-steel

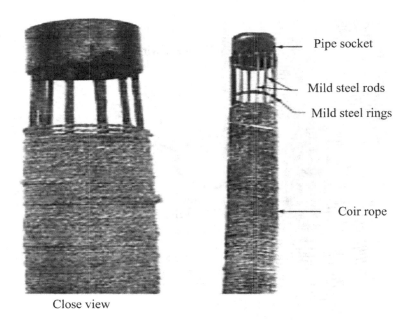

Close view

FIGURE 11.3   Coir-rope strainer with mild-steel rod frame.

longitudinal rods or bars. Iron rings with a cross-section of 1 cm × 3 mm are used to reinforce the frame at appropriate intervals. The longitudinal rods' one end is riveted or welded to threaded mild-steel pipe sockets (Figure 11.3). The other end is welded to a threaded pipe ring that fits into the socket end of the previous screen section. As a result, each length of the strainer can be connected to the next. The entire surface of the coir acts as a screen, allowing water to enter while preventing sand from entering. The main disadvantage of a coir strainer is that it has a limited lifespan of about 3–5 years. The rusting of the iron bars/strips of the supporting frame is the primary cause of its failure. This reduces the coir string's strength at the contact points. The strainer's life is extended by coating the iron frame with bitumen. Another reason for failure is loosening of the coir string, which expands when wet. Winding the string at an appropriate tension extends the life of the coir screen. When synthetic fiber rope (nylon rope) is used in place of coir, the strainer's durability is greatly increased.

The coir-rope strainer cannot withstand the high pressure created by an air compressor during the development of tube wells. As a result, such wells can only be developed using pumps.

Despite these limitations, the coir strainer is widely used in shallow tube wells in deltaic regions of India due to its low cost.

11.6   BAMBOO STRAINER

The bamboo strainer is similar to a coir-rope strainer, except that the longitudinal members of the supporting frame are made of bamboo strips rather than steel rods

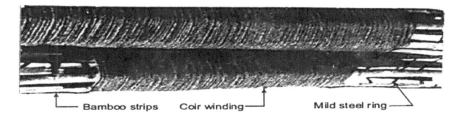

FIGURE 11.4   Coir-rope strainer with bamboo frame.

(Figure 11.4). Bamboo strips are laid lengthwise and secured on mild-steel rings 10–12 cm in diameter placed at approximately 30 cm intervals to make the strainer. The cylindrical frame is wrapped in coir rope. Bamboo strainers, which were invented in Bihar, have gained popularity as a low-cost material for shallow tube wells, particularly in the Gangetic plains of Bihar and West Bengal. Bamboo strainers have the same adaptability as regular coir strainers. Microorganisms such as fungi and termites, on the other hand, significantly reduce the durability of bamboo strainers. The lifespan of a bamboo strainer can be extended by coating it with epoxy, tar, or bituminous paint. The average lifespan of a bamboo strainer is 5–6 years (Jha, 2004).

11.7   AGRICULTURAL STRAINERS

Agricultural strainers come in a variety of shapes and sizes. A perforated galvanized iron pipe is welded to iron hoops or strips of 1 cm width and 3 mm thickness in the most typical style of agricultural strainer. The perforations are circular and are drilled out using a drill. They're positioned in such a way that the perforated sections aren't covered. The iron hoops and strips are wrapped in copper netting and soldered (Figure 11.5). The type of copper mesh employed is determined by the sand layers encountered during the bore. In most cases, the market offers three varieties of copper nettings: fine, medium, and coarse. The wire netting is not in direct touch with the perforated tube but it is wrapped around in this arrangement. As a result, the copper netting in front of the holes has no effect on the perforation area. The above configuration improves hydraulic efficiency. The agricultural strainer has about the same versatility as the coir strainer. It might, however, be utilized for wells that are significantly deeper. Perforated castings, pipes composed of different materials, such as ceramics and asbestos cement, have been employed. However, they have the drawback of being extremely heavy and difficult to install.

Furthermore, such screens have a small percentage of open space. Cage-style screens are sometimes utilized. Instead of perforated pipe, the outer jacket is mounted on a frame of longitudinal strips reinforced at intervals with rings/hoops. However, due to its low strength, this design has a very restricted use. A pipe-base well screen, which consists of a slotted PVC pipe with vertical slots, is the strainer that corresponds to the agricultural strainer in industrialized countries. On the perforated pipe, a continuous-slot screen jacket is installed. Winding triangular PVC wire directly on the slotted pipe creates the jacket. Pipe sockets are placed at the ends of the screen.

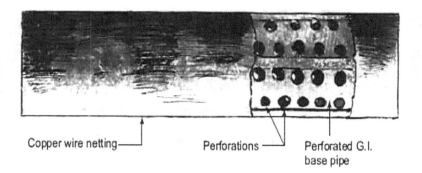

Copper wire netting———   Perforations ———   Perforated G.I.
                                                    base pipe

FIGURE 11.5   Agricultural strainer.

11.8   SLOTTED-PIPE WELL SCREENS

Slotted-pipe well screens are typically utilized with a gravel pack around them to
prevent fine sand from entering the well. In the event of aquifer formations with a
mixture of fine and coarse particles, or in the case of comparatively deeper wells,
such screens are used. There are several types of slotted-pipe screens (Figure 11.6)
in the market.

11.9   MILD-STEEL SLOTTED-PIPE WELL SCREENS

Mild-steel slotted-pipe well screens (Figure 11.6) are often constructed from
medium- and heavy-duty mild-steel pipes. The slit should be shaped and sized in
such a way that the aquifer material or gravel does not block the open region of the
screen. The entire open area ranges from 15% to 22% of the pipe's surface area.
The size of the slot is set based on the sieve analysis of the aquifer material, such
that the finer fractions of the formation are removed during the tube well development stage and the coarser fractions are kept surrounding the well screen. The
slots should not be too wide to allow pebbles into the screen, causing it to become
clogged. The most frequent slot sizes in India are 1.6 and 3.2 mm wide by 10 cm
long. The minimum distance between slots is 3 mm. The slots are organized in
such a way that the flow is evenly distributed throughout the screen's perimeter. To
provide enough strength to the well screen, they are distributed in groups of three
or four and organized so that the slots of one group are not aligned with those of the
neighboring row. Figure 11.6 depicts a typical view of the slot configuration. Both
ends of the slotted pipe are threaded. A blind pipe 1.25 m long with a cap called
a bail plug is installed at the bottom end of the slotted pipe. In the event that the
bail plug fails, an 'eye' is attached within to aid in the extraction of the tube well
assembly.

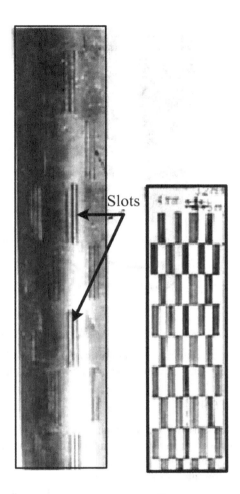

FIGURE 11.6 Multi steel slotted-pipe well screen.

11.10 PRE-PACK FILTERS

The gravel pack in pre-pack filters is cast on slotted pipes. Graded hard gravel grains are coated and glued together with waterproof chemicals on the outside surface of a slotted pipe in the production of pre-pack filters. Because the gravel is loosely fixed, it has a high permeability value. The gravel size utilized varies based on the aquifer's grain size dispersion. Pre-pack filters are suited for shallow and medium-deep tube wells in locations with fine sand, coarse sand, and gravel aquifers. Pre-packing,

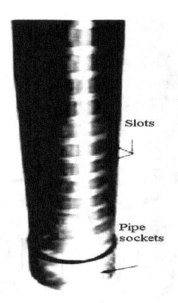

FIGURE 11.7   Brass screen with slots.

it is said, saves money by reducing the amount of gravel used, protects the gravel envelope from dislodging, and increases the collapse and tensile strength greatly.

11.11   BRASS SCREENS

The most expensive well screens are brass screens (Figure 11.7) and stainless steel screens. Brass screens are commonly used in drinking water tube wells. Such screens are favored over mild-steel slotted pipes because they last longer, are resistant to incrustation and corrosion, and prevent water pollution from rusting. They're made of brass/stainless steel sheets that are around 4 mm thick. Copper, zinc, and lead are used to make brass sheets. Copper, zinc, and lead have a 60%, 38.75%, and 1.25% ratio, respectively. The slots are made to match the well screen's design. The slotted sheet is rolled into a tube and welded together. Outer threads are given on the extremities of the tube lengths (unslotted) for attaching the tubes through a socket. Commercially available screens come in a variety of slot widths and diameters, and the size required is determined by the aquifer's design and particle size distribution.

REFERENCES

Boman, B., Shukla, S. and Hardin, J. D. 2006. "Design and Construction of Screened Wells for Agricultural Irrigation Systems". Circular 1454, University of Florida, Gainesville, FL, 12.

Jha, U. M., 2004. "Economics of bamboo boring: A study of north-east region of Bihar", *Project Report Sponsored by Planning Commission (SER Div.), Government of India,* Tilkamanjhi Bhagalpur University, Bhagalpur, p. 93.

Johnson, Edward E., 1966, *"Ground Water and Wells"*, Edward E. Johnson Inc., Saint Paul, Minnesota, USA, p. 440.

12 Study of Different Drilling Equipment and Drilling of a Tube Well

Objective: Study of Different Drilling Equipment and Drilling of a Tube Well

12.1 THEORY

Tube well drilling rigs come in a variety of shapes and sizes to accommodate the wide range of subsurface deposits. Water well drilling equipment is divided into three categories:

    i. Percussion (cable tool) drills,
    ii. Rotary drills (including direct rotary, reverse rotary, and air rotary), and
    iii. Down-the-hole (DTH) hammer drills are the three types of drills available.

Since the twelfth century A.D., the cable tool drilling technology has been used to drill water wells. Drilling activities are carried out through this technique by lifting and dropping a heavy string of tools into the borehole on a regular basis. In the hole, no fluid is circulated. Drill cuttings are bailed off by adding water after the drill bit splits or smashes the formation into small bits (Das, 1983).

The rotary method involves cutting the formation with a rotating bit and lifting the cutting with the circulation of a drilling fluid as the bit penetrates. The DTH drilling method combines percussion drilling's percussive action with rotary drilling's rotational motion. A pneumatic hammer is attached to the end of a drill string to create the tool. Drilling rigs are classified into three types based on their capacity ratings: shallow, medium, and heavy. They can be further categorized under each group based on the diameter and depth of the hole that can be drilled with the equipment. A popular classification of drilling rigs is shown in Table 12.1.

12.2 DRILLING METHODS

The type of drilling equipment used to construct the tube well is used to classify drilling procedures. The following is a list of the most widely utilized drilling methods:

12.2.1 PERCUSSION DRILLING

    a. Hand boring.
    b. Mechanical boring (standard cable tool method).

DOI: 10.1201/9781003319757-16

TABLE 12.1
Classification of Drilling Rigs

Type of Drilling Rig	Classification	Limiting Diameter of Hole (mm)	Depth of Hole (m)	Size of Drill Rods (mm)
Percussion (cable tool)	Light	130	<50	–
	Medium	200	50–170	–
	Heavy	200	>170	–
Direct rotary	Light	200	0–250	73
	Medium	200	251–457	73–89
	Heavy	200	>457	89
Reverse rotary	Light	500	<170	150
	Medium	675	>170	150
Combination (rotary-cum-percussion)	Light	–	–	–
	Medium	200	<500	89
	Heavy	300	>500	89
Down-the-hole (DTH) hammer	Light	114	<50	89
	Medium	150	50–170	114
	Heavy	200	>170	114

Source: Das (1983).

12.2.2 HYDRAULIC ROTARY DRILLING

a. Direct circulation hydraulic rotary drilling.
b. Reverse circulation hydraulic rotary drilling.
c. Dual wall reverse circulation hydraulic rotary drilling.

12.2.3 MISCELLANEOUS DRILLING METHODS

a. Jetting.
b. Core drilling.
c. Calyx drilling.
d. Sonic drilling.

12.3 PERCUSSION DRILLING

The percussive and cutting operations of a drill bit that is raised and dropped alternately form the well bore in percussion drilling. The operation shatters subterranean formations into shards. The loosened material is mixed into sludge by the reciprocating motion of the drilling instruments, which is evacuated from the hole at intervals by bailer or sand pump. Simultaneously, casing pipes are lowered to prevent material from cave-in and to allow drilling to continue. Both manual labor (hand boring) and power rigs can be employed for percussion drilling (standard cable tool method).

12.4 HAND BORING

In unconsolidated subsurface formations, hand drilling is often used in the building of shallow tube wells.

The hand boring set (Figure 12.1) is a simple, manually driven device that works on the basis of percussion. A tripod, crab winch, bailers or sand pumps, cutter shoes, sinking pipe loading system, chain wrenches, pulley blocks, flexible steel wire rope, and tools such as spanners and wrenches are included in the equipment. Normally, the bailer/sand pump is used as a drilling instrument. It is a steel pipe with a cutting edge at the lower end and a hook at the upper end, with a thickness of around 6 mm. Inside the bailer, right above the cutting sleeve, is a flap valve. When boring, a winch aids in the winding and unwinding of the wire rope. The boring pipes are made of mild steel and have appropriate joints.

When joining bore pipes, three types of joints are commonly used: (i) flush joints, (ii) swollen and crossed joints, and (iii) socketed joints. Casing pipes with flush joints are easy to lower and withdraw, but for the same strength, thicker pipes are required. A cutter shoe is installed at the bottom of the casing pipe, which is somewhat larger in diameter than the casing pipe itself. This makes the installation of boring pipes and their subsequent removal after boring and the installation of well casing pipes and screens easier.

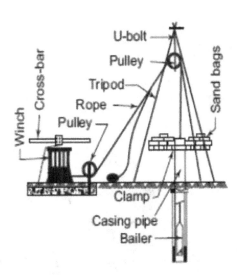

FIGURE 12.1 Schematic view of a manually operated percussion drill for constructing shallow tube well.

12.5   HAND BORING PROCEDURE

Figure 12.2 shows a schematic drawing of the hand boring method of well construction. At the site, a pit with a diameter of 2–2.5 m and a depth of 2 m has been dug. An auger is used to drill a hole 1 m deep and slightly larger in diameter than the bore. With the help of a tripod, a boring pipe is lowered into the bore hole. After partially lowering the first pipe into the hole, it is clamped in place and loaded with sand bags (Figure 12.2). After that, it is partially filled with water. Boring is continued by manipulating the bailer, which is suspended from a tripod stand by a wire rope going through a pulley. To prevent the tripod legs from slipping or tilting, they should be buried in the ground. The wire rope that passes through the pulley is attached to a winch and can be coiled or unwound as needed to lower and raise the bailer.

The bailer is positioned centrally over the drilling pipe due to the tripod's rigidity. In the boring pipe, the bailer is moved vertically up and down. From the top, water is put into the boring pipe until it reaches the water table. The torsion in the wire rope automatically imparts a tiny circular motion to the bailer. The bailer's flap valve is forced open during the downward stroke, and the loose material hammered at the bottom of the bore enters the bailer. The flap valve is closed and the material inside is retained when the upward stroke begins. The bailer is raised out of the drilling pipe by twisting the rope on the winch after roughly 30–40 strokes, and the loose material is discharged. The material that the bailer brings up is a sample of the stratum that was encountered during drilling. Boring will continue and additional casing pipes will be lowered until the well reaches the appropriate depth.

FIGURE 12.2   Loading of sand bags on wooden clamps for lowering casing pipes in a manually drilled tube well.

12.6 MECHANICAL PERCUSSION BORING

Cable tool drilling refers to mechanical percussion methods of drilling tube wells. Drilling is done in the same way as hand boring is done: by moving the tools up and down. The hard formations are broken or crushed into little bits by the drill bit. The drill bit loosens the material when operating in soft, unconsolidated rocks. The reciprocating movement of the instruments in both circumstances creates a slurry or sludge by mixing the crushed or loosened particles with water. If no water is available in the strata being penetrated, the appropriate water is introduced to the borehole to generate the slurry. A sand pump or bailer is used to drain the resultant slurry from the borehole at regular intervals. Sludge builds up on the tools, slowing their movement and, as a result, their penetration. The frequency with which the sludge is pumped from the well is usually determined by this component.

12.7 EQUIPMENT FOR CABLE TOOL PERCUSSION DRILLING

A portable rig (Figure 12.3) installed on a truck chassis or on a trailer is conventional equipment for cable tool drilling. The basic components of a percussion rig are (i) a mast (derrick), (ii) a two- or three-line hoist, (iii) a spudder for raising and lowering the tools, and (iv) a diesel engine to power the numerous activities (Figure 12.4). One line (the drilling reel) is used to operate drilling instruments, the second (the sand reel) is used to operate bailer/sand pump, and the third (the casing reel) is used to operate boring pipes in a three-line hoist. The mast is designed long enough to hoist the longest string of tools/pipes out of the pit, with a minimum height of roughly 10 m. During transit, the mast is telescoped to its shortest length and folded at the machine's top. The reels must be large enough to store enough cables for the deepest holes to be drilled. The rig can drill and is sturdy enough to handle the drilling and

FIGURE 12.3 A truck mounted heavy duty percussion drilling rig in transport position.

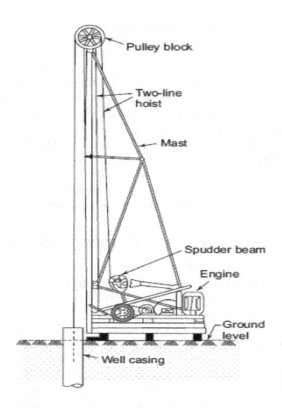

FIGURE 12.4   Basic components of percussion drilling rig.

fishing tools, as well as other gear. The drilling process consists of three steps: drilling the hole, bailing out the drilled cuttings, and capping the bored hole. These three procedures are usually carried out one after the other until the bore is completed (Garden, 1958).

12.8   HYDRAULIC ROTARY DRILLING

Cutting a borehole with a rotating drill bit and removing the cuttings with the continuous circulation of a drilling fluid is what hydraulic rotary drilling is all about. The drill bit is affixed to the bottom of a drill pipe string. The drill bit and the drilling fluid are the two most critical components of the rotary drilling method. Both are required for the cutting and maintenance of the borehole. The rotary drilling machine's components are all built to perform two functions at the same time: bit operation and continuous drilling fluid circulation. Direct circulation drilling and reverse circulation drilling are the two methods of hydraulic rotary drilling (Figure 12.5).

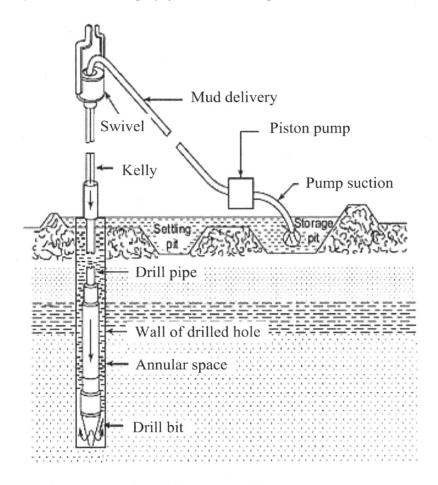

FIGURE 12.5   Basic principles of direct rotary drilling.

12.9 DRILLING WITH DTH HAMMER AND AIR ROTARY DRILLS

Due to the sluggish penetration rate, high bit cost, and high machinery maintenance cost, percussion and rotary well drilling methods are frequently uneconomical in water well drilling in hard rock locations. For drilling water wells in hard rock locations, air-operated DTH drills and air rotary drills have proven to be the most effective. Drilling with DTH machines blends percussion drilling's percussive action with rotary drilling's rotational motion. A pneumatic hammer is attached to the end of a drill string to create the tool. The hammer tool is made up of an air-operated piston that strikes the bit at a high frequency while being hydraulically forced into the rock. A hammering force is transmitted to the complete surface of the bottom of the hole by spinning the drill string at the same time, pulverizing the formation. The air entered for the hammering action is allowed to escape through the annular space between the hole and the drill string at the bottom of the hole, lifting the cuttings to the surface. Because the hammer at the bottom of the hole allows the piston to deliver the blows directly to the bit without losing energy through the drill string, DTH drilling provides for faster penetration. The hammer can drill granite up to three times faster than an air rotary drill and several times faster than percussion drilling. The bit's operation is influenced by the air compressor's capacity, the hammer's design, the bit's kind, and the type of coolant used throughout the drilling process. An air compressor supplies the necessary amount of air to run the hammer and clean the cuttings on a constant basis. Short, fast blows assure a straight hole by reducing the effect of chipping and fracturing the formation. DTH drilling, on the other hand, is limited to small-diameter wells. The most common drill bit size is 15 cm. When rock-roller bits are utilized, however, the well size can be bigger (approximately 30 cm). Drilling in rocks with significant air pockets and deep water columns presents challenges. Air velocity in the annular space should be maintained between 600 and 1,500 m/min to properly remove the cuttings. A velocity of roughly 900 m/min in the annular space is usually recommended. In some formations, excessive air velocities can produce hole enlargements. Surface equipment with relatively low back pressures can greatly reduce the annular air velocity near the surface.

12.10   AIR ROTARY DRILLING

Instead of drilling mud, compressed air is used as the drilling fluid in air rotary drilling. Air is circulated via the drill pipe in this form of well drilling. It exits through drill bit pores and rises in the annular area, carrying drill cuttings to the surface or blowing them out into rock crevices. Only consolidated materials are suitable for air rotary drilling. A standard mud pump and a high-capacity air compressor are generally included in air rotary drilling equipment. Drilling mud can be used to drill through caving materials that are higher than the bedrock. When drilling with air, roller-type rock bits, similar to those developed for mud fluid drilling, can be utilized. Three-cone rock pieces with a diameter of around 30 cm are widely used. When compared to the traditional hydraulic rotary method, the air rotary method usually results in faster penetration in hard rock formations. Drilling is not hampered by a lack of circulation. It is feasible to sample strata accurately. The primary stumbling block is the high amount of air required for drilling. Taking up cave formations presents its own set of challenges. As a result, pneumatic direct rotary drilling has primarily been utilized to drill small-diameter holes in semi-consolidated strata where hydraulic direct rotary drilling would be too sluggish and a DTH hammer would be too dangerous.

12.11   MISCELLANEOUS DRILLING METHODS

The basic drilling technologies, such as percussive, hydraulic rotary, DTH, and air rotary systems, were covered in previous sections. Furthermore, diverse drilling applications necessitate distinct drilling equipment to meet specific site constraints and/or cost. Core drilling, calyx drilling, jetting, and sonic drilling are among them (Michael et al., 2008).

12.12   CORE DRILLING

Core drilling's major goal is to gather uncontaminated samples of subsurface formations for a variety of applications, including soil sampling, geological and mining investigations, groundwater prospecting, and foundation engineering. Soil and rock sampling, as well as the planning and inspection of civil engineering projects, are common uses for diamond core drills. They are also used to drill blasting holes and inject grout into them.

12.13   CALYX DRILLING

Calyx drilling is commonly utilized in hard rock formations for core drilling and the construction of large diameter shallow tube wells. It is less pricey in comparison. A calyx drill bit is made up of a hollow steel tube with two inclined slots that is attached beneath another tube (core barrel) that is connected to the drill rods. These are mechanically rotated. Through the drill string, chilled shots and water are supplied to the bottom of the bit. The shot bit grinds these into abrasive material with sharp edges that cut into the solidified formation, generating an annular ring that serves as a core inside the core barrel. By grouting the core barrel with quartz chips, the core barrel is removed from the well.

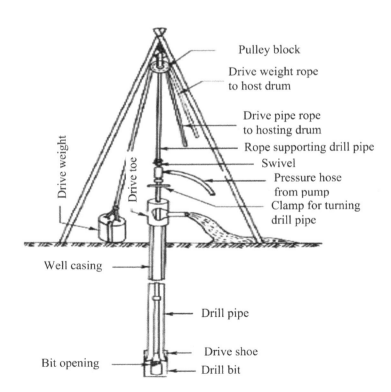

FIGURE 12.6   A simple jetting rig.

12.14 JETTING

The jetting method of well construction is primarily limited to alluvial formations with a shallow water table. The approach is popular in West Bengal and Orissa, India. Jetting is useful for boring through stubborn, tenacious clays, where conventional percussion rigs typically take a long time to bore. A drill bit with nozzles is attached to the drill pipes at the bottom of a jetted well. Water is pumped through the bit at a high pressure. The drill rods are manually turned. A high-velocity stream of water is used to drill the hole. As it strikes the material, the stream loosens it and washes the finer particles, carrying them upward and out of the hole. Construction of a jetted well can be done in a variety of ways depending on the equipment available (Figure 12.6).

REFERENCES

Das, S. K. 1983. Key note address in tech. session on drilling and drilling equipment for ground water extraction. *Paper presented in II All India Conf. on Drilling and Drilling Equipments*, Institute of Public Health Engrs., New Delhi, India, Dec. 26–28, 13.

Garden, R. W. 1958. *Water Well Drilling with Cable Tools*. Bucyrus-Erie Co., South Melwaukie, WI, 230.

Michael, A.M., Khepar, S.D. and Sondhi, S.K. 2008. Water Wells and Pumps. Tata McGraw-Hill Publishing Company Limited, New Delhi, India, 1–52.

13 Measurement of Water Level and Drawdown in Pumped Wells

Objective: Measurement of Water Level and Drawdown in Pumped Wells

A pumping test is a field experiment in which a well is pumped at a controlled rate and the water-level response (drawdown) is measured in one or more surrounding observation wells and, if desired, in the pumped well (control well); response data from pumping tests are used to estimate the hydraulic properties of aquifers, evaluate well performance, and identify aquifer boundaries. Pumping tests are sometimes known as aquifer tests or aquifer performance tests. A drawdown test is a type of pumping test.

13.1 PURPOSE OF PUMPING TEST

The pumping test, also known as an aquifer test, is the sole method for measuring the hydraulic properties of various aquifer systems [e.g., T, K, S, or S_y, leakage factor (B), and hydraulic resistance (C)] as well as those of producing wells that are currently accessible (e.g., well parameters, safe well yield). Long-term time-drawdown pumping tests can also reveal the presence, if any, of subsurface hydraulic barriers, as well as the sort of boundary that exists (recharge, impermeable, or leaky boundary). The degree of heterogeneity and anisotropy in aquifer systems can be determined by strategically placing observation wells in various locations and directions. Step-drawdown tests can also reveal vital details regarding the hydraulic features of production wells (such as the aquifer loss coefficient, well loss coefficient, well efficiency, well-specific capacity, and safe well yield) as well as their condition.

13.2 TYPES OF PUMPING TESTS

There are primarily four types of pumping tests/aquifer tests. They are as follows:

1. Time-Drawdown Test:
   a. Interference Test
   b. Distance-Drawdown Test
   c. Single-Well Test
2. Recovery Test
3. Step-Drawdown Test or Variable Rate Test
4. Injection Tests:
   a. Time-Groundwater Level Rise Test
   b. Step-Injection Test

DOI: 10.1201/9781003319757-17

13.3 TESTS FOR DETERMINING HYDRAULIC PARAMETERS OF AQUIFERS

13.3.1 Steady and Unsteady Time-Drawdown Tests

Aquifer parameters can be determined using time-drawdown pumping tests. Time-drawdown pumping tests are classified into two types based on the type of data produced: (i) unsteady-state tests or transient tests and (ii) steady-state tests or equilibrium tests. The pumping rate is kept constant in both categories throughout the duration of the test. Groundwater level changes in response to a constant pumping rate are measured over time in unsteady-state tests. The results of transient tests can be used to calculate almost all of the hydraulic parameters of aquifer systems.

Pumping is continued in steady-state tests until near-equilibrium conditions are reached (i.e., there are negligible changes in groundwater levels with time). In most aquifer systems, steady-state tests only approach quasi-steady-state conditions. Because of the constant aquifer recharge and discharge in a groundwater system, true equilibrium may never be achieved under field conditions. The results of steady-state tests can only be used to calculate aquifer transmissivity or hydraulic conductivity and, in some cases, leakance (leakage coefficient); the storage coefficient of an aquifer system cannot be calculated using these results.

While conducting time-drawdown tests, drawdown can be measured either in one or more observation wells (called "interference test") or in the pumping well itself (called "single-well test"). They are described below (Franklin and Zhang, 2002).

13.3.2 Interference Test

An interference measurement is the measurement of groundwater level changes (drawdowns) in response to pumping in one or more observation wells, as shown in Figure 13.1. The interference test can be steady-state or unsteady-state, depending on the nature of the interference. Because interference measurements, unlike pumping well measurements, do not include turbulent flow components, the drawdown measured in observation wells is representative of that in aquifers. This is why interference-test data are preferred for calculating aquifer parameters. It is critical for interference tests that the observation wells be completed in the same aquifer as the pumping well.

Aquifer transmissivity (T) or hydraulic conductivity (K) determination from steady-state test data typically requires steady drawdown data from two observation wells or steady drawdown data from at least one observation well in addition to the pumping well; drawdown measured in a pumping well must be corrected for well losses. However, if the radius of influence is known, steady drawdown data from a single observation well or pumping well can also be used to compute K or T. On the other hand, determining aquifer parameters such as storativity (storage coefficient) requires only transient drawdown data from a single observation well.

13.3.4 Distance-Drawdown Test

Drawdown is measured in three or more observation wells located at different radial distances from a pumping well in distance-drawdown tests, yielding a set of

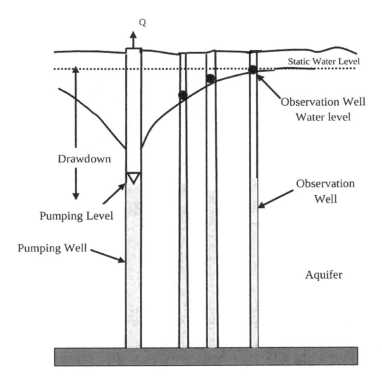

FIGURE 13.1   Illustration of an interference pumping test.

distance-drawdown data. This test is not performed separately in the field; rather, such a dataset can be created from time-drawdown measurements taken at various locations during a time-drawdown test. Thus, depending on the nature of the time-drawdown tests, distance-drawdown data can be of two types: distance-steady drawdown data or distance-unsteady drawdown data; the former dataset can only yield aquifer transmissivity (T) or hydraulic conductivity (K), whereas the latter dataset can yield both T and storage coefficient (S). Distance-drawdown data can be used to cross-check aquifer parameters derived from time-drawdown data.

13.3.5   Single-Well Test

A single-well test is performed when the time-drawdown aquifer test is carried out in such a way that the drawdown is measured only in a pumping/production well. When there are no observation wells in a basin or sub-basin and it is not possible to install observation wells due to financial and/or time constraints, a single-well test is normally performed. As a result, single-well tests are less expensive than interference tests.

It should be noted that the drawdown in a pumping well is made up of two types of head losses: (i) aquifer loss or formation loss (drawdown due to laminar flow in the aquifer, also known as 'theoretical drawdown') and (ii) well loss (drawdown due

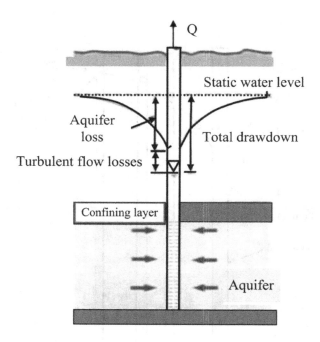

FIGURE 13.2 Components of drawdown in a pumping test.

to turbulent flow in the immediate vicinity of the pumping well through the screen and/or gravel pack, as well as flow inside the well to the pump intake), as shown in Figure 13.2. Because the well loss is caused by turbulent flow, it is proportional to the nth power of the pumping rate (well discharge), where n is a constant greater than one. As a result, the total drawdown in a pumping well, also known as well drawdown (s_w), is given as (Figure 13.2):

$$s_w = \text{Aquifer Loss} + \text{Well Loss}$$

or

$$s_w = BQ + CQ^n$$

where

   B: aquifer/formation loss coefficient,
   Q: pumping rate (well discharge), and
   C: well loss coefficient which is a function of the radius, construction, and condition of the pumping well.

13.3.6 Recovery Test

A recovery test is performed at the conclusion of a time-drawdown aquifer test. It is an unsteady-state aquifer test in which groundwater rise is measured over time in a

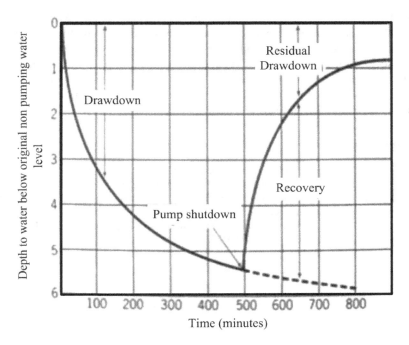

FIGURE 13.3   Illustration of time-drawdown and recovery tests.

pumping/production well or an observation well after pumping has ceased (Figure 13.3). As shown in Figure 13.3, when a well's pumping is stopped, the groundwater level in the aquifer gradually rises and should theoretically return to its pre-pumping level; this time period is known as the recovery period or recovery phase. At various times during the recovery period, the 'residual drawdown' (difference between the pre-pumping groundwater level and the depth of groundwater after the pump is turned off) or 'recovery' (difference between the extrapolated time-drawdown curve and the residual drawdown) is computed from the water-level data measured during the recovery period (Figure 13.3). The resulting 'time-residual drawdown' or 'time-recovery' data are used to calculate aquifer parameters.

Because of the lack of water-level fluctuations caused by discharge variations and the absence of turbulence, data obtained during the recovery period are more reliable than data obtained during the pumping period. Time-recovery data measured in pumping wells, like time-drawdown data from single-well tests, can only yield T or K. T and S can, however, be determined if the time-recovery data are measured in observation wells.

13.3.7   INJECTION TESTS

Injection tests can be carried out in areas where artificial recharge projects using the well injection technique are in operation. The injection well test procedures and equations are similar to those used for drawdown tests, except that the injection pressure head (the difference between the static groundwater level and the groundwater level during injection) is substituted for drawdown and the injection rate for pumping rate (Roscoe Moss Company, 1990). As a result, the data from the time-groundwater level rise test can be used to calculate the hydraulic parameters of aquifer systems. Similarly, step-injection test data can be used to determine the hydraulic properties of production wells (Jha and Sarma, 2020).

REFERENCES

Franklin, W. and Zhang, H. 2002. Fundamentals of Ground Water. https://www.wiley.com/en-us/Fundamentals+of+Ground+Water-p-9780471137856
Jha, M. and Sarma, P. 2020. Groundwater, Wells and Pumps. https://agrimoon.com/wp-content/uploads/Groundwater-Wells-and-Pumps.pdf
Roscoe Moss Company. 1990. Handbook of Groundwater Development. https://onlinelibrary.wiley.com/doi/book/10.1002/9780470172797

14 Well Design under Confined and Unconfined Conditions

Objective: Well Design under Confined and Unconfined Conditions

14.1 STEADY-STATE FLOW

The flow is said to be steady when no change occurs with time, i.e.

$$\frac{dv}{dt} = 0$$

where
  v: Velocity of flow, m/s and
  t: Time, s

When there is an equilibrium between the discharge of the pumped well and the recharge of the aquifer by an outside source, steady-state flow occurs.

The flow conditions in unconfined and confined aquifers differ and must be considered separately (Subramanya, 2013).

14.2 STEADY-STATE FLOW TO WELLS IN UNCONFINED AQUIFERS

An equation for steady radial flow to a well in an unconfined aquifer can be derived with the Dupuit assumptions which state that:

1. The velocity of flow is proportional to the tangent of the hydraulic gradient.
2. The flow is horizontal and uniform everywhere in a vertical section.
3. The flow is assumed two dimensional to a well centered on a circular land and penetrating a homogeneous and isotropic aquifer.

As shown in Figure 14.1, the well completely penetrates the aquifer to the horizontal base. The well discharge Q at any distance r is expressed as

$$Q = Kia = 2\pi r h K \frac{dh}{dr}$$

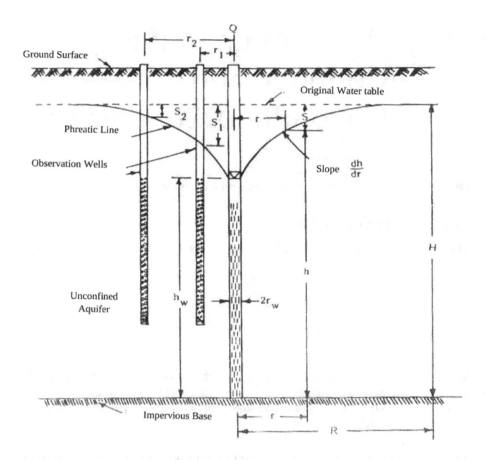

FIGURE 14.1 Steady-state flow into a fully penetrating well in an unconfined aquifer.

or

$$hdh = \frac{Q}{2\pi K} \frac{dr}{r}$$

For steady radial flow to the well, integrating for boundary conditions at the well:
$h = h_w$ at $r = r_w$ and
$h = H$ at $r = R$

$$\int_{h_w}^{H} h \, dh = \frac{Q}{2\pi K} \int_{r_w}^{R} \frac{dr}{r}$$

or

$$H^2 - h_w^2 = \frac{Q}{\pi K} \ln(R / r_w)$$

$$Q = \frac{\pi K (H^2 - h_w^2)}{\ln(R/r_w)}$$

$$= \frac{\pi K (H^2 - h_w^2)(H - h_w)}{\ln(R/r_w)}$$

(14.1)

where

Q: Constant discharge rate, m³/s
H: Original elevation of the water surface, measured from the impervious base, m
h_w: Depth of water in the well, measured from the impervious base, m
R: Radius of influence of the well, m
K: Hydraulic conductivity of the aquifer, m/s and
r_w: Radius of well, m

Thiem (1870) demonstrated the applicability of Eq. (14.1). He demonstrated that the drawdown of the phreatic surface face from the original groundwater table became negligible beyond a certain distance from the well. The above-mentioned Dupuit–Thiem theory is critical in well hydraulics (Raghunath, 2006).

14.3   EVALUATION OF HYDRAULIC PROPERTIES

The hydraulic properties of the aquifer can be evaluated by using Eq. (14.1) for steady-state conditions.

Let the steady-state drawdown at the observation wells be s_1 and s_2 and r_1 and r_2 the distance of the observation wells from the center of the test well (Figure 14.1).

Since $h = H - s$, then Equation (4.4.1) can be transformed as

$$Q = \frac{\pi K (h_2^2 - h_1^2)}{\ln(r_2 / r_1)}$$

which can be expanded into

$$Q = \frac{\pi K [(H - s_2)^2 - (H - s_1)^2] 2H / 2H}{\ln(r_2 / r_1)}$$

in which h_1 and h_2 are elevations of water surface, measured from impervious base at observation wells 1 and 2.

Replacing $s - s^2/2H$ by $s' = $ the corrected drawdown yields

$$Q = \frac{2\pi K H (s_1' - s_2')}{\ln(r_2 / r_1)}$$

(14.2)

in which

$$s_1' = s_1 - s_1^2 / 2H$$

$$s_2' = s_2 - s_2^2 / 2H$$

s_1' and s_2' are corrected steady-state drawdowns at points 1 and 2, respectively, or

$$Q = \frac{2\pi T(s_1' - s_2')}{\ln(r_2 / r_1)}$$

in which

$$T = KH = \text{The assumed transmissibility of the aquifer, m}^2/\text{s}$$

or

$$T = \frac{Q\ln(r_2 / r_1)}{2\pi(s_1' - s_2')} \tag{14.3}$$

The values of the transmissibility and hydraulic conductivity can be estimated using Eq. (14.3) only when the drawdown in the aquifer is small in relation to the thickness of the saturated portion of the aquifer.

14.4   STEADY-STATE FLOW TO WELLS IN CONFINED AQUIFER

Dupuit used an equation to derive the radial flow equation for a well completely penetrating a confined aquifer. The flow is assumed to be two dimensional to a well centered on a circular island that penetrates a homogeneous and isotropic aquifer. Because the flow is horizontal everywhere, the Dupuit assumptions are correct. Using plane polar coordinates and the well as the origin, the well discharge Q at any distance r when the aquifer thickness b is calculated as follows:

$$Q = Kia = 2\pi \, rbK \frac{dh}{dr}$$

or

$$dh = \frac{Q}{2\pi \, Kb} \frac{dr}{r}$$

Integrating for the boundary conditions at the well, $h = h_w$ at $r = r_w$ and $h = H$ at $r = R$ at the extremity of the area of influence,

$$\int_{h_w}^{H} dh = \frac{Q}{2\pi Kb} \int_{r_w}^{R} \frac{dr}{r}$$

$$H - h_w = \frac{Q}{2\pi Kb} \ln(R / r_w)$$

After rearranging,

$$Q = \frac{2\pi Kb(H - h_w)}{\ln(R / r_w)} \tag{14.4}$$

where

    b: the thickness of the horizontal pervious stratum confined between two horizontal impervious strata, m

The other variables are the same as defined in Eq. (14.2). Equation (14.4) can be used to evaluate the hydraulic properties of an aquifer, based on the measurements made during the test. Thiem (1870), who worked independently of Dupuit, derived Eqs. (14.2) and (14.4), based on the following assumptions which were more precisely defined than those of Dupuit:

1. The aquifer has a seemingly infinite areal extent.
2. The aquifer is homogenous, isotropic, and of uniform thickness over the area influenced by the pumping test.
3. The pumped well penetrates the entire thickness of the aquifer and receives water from its entire thickness by horizontal flow.
4. Flow to the well is in steady state.

14.5 EVALUATION OF HYDRAULIC PROPERTIES

In the case of wells in a confined aquifer, either of the following two procedures can be used to determine the hydraulic properties of the water-bearing formations (Michael *et al.*, 2008).

14.5.1 PROCEDURE I

The observed drawdown in each piezometer or observation well is plotted against the corresponding time on a single logarithmic paper, with drawdown on the vertical axis, on a linear scale, and time on the logarithmic scale. Each piezometer's time drawdown curve that best fits the points is downward. The curve of the different piezometers that best fits the points will be drawn. For the latter time data, the curves of these piezometers run parallel, and thus the mutual distance is constant. This implies that the hydraulic gradient is constant and that the flow in the aquifer is in a steady state. To solve for the transmissibility, the values of the steady-state drawdown of two piezometers are substituted in Eq. (14.4) along with the corresponding values of r and the known values Q.

$$T = Kb$$

$$Q = \frac{2\pi(s_1 - s_2)}{\ln(r_2 / r_1)} \tag{14.5}$$

where s_1 and s_2 are the values of drawdown of the piezometers and r_1 and r_2 distances from the center of the well, respectively.

    To obtain a more precise value of T, the same process should be repeated for all possible piezometer combinations. The results should, in theory, show a high degree of agreement. However, because the results are usually slightly different, an average value can be used.

14.5.2 PROCEDURE 2

On a semi-logarithmic paper, the observed steady-state drawdowns of each observation well are plotted against the distance between the pumping well and the piezometer. The distance is plotted on a logarithmic scale on the horizontal axis, and the drawdown is plotted on a linear scale on the vertical axis. The distance-drawdown curve is formed by drawing the best-fitting straight line through the plotted points (Figure 14.2). The slope of the distance-drawdown curve is determined for the logarithmic cycle of distance, s. When the value of s is substituted in Eq. (14.5), the following relationship is obtained:

$$Q = \frac{2\pi T \Delta s}{2.30} \tag{14.6}$$

The values of transmissibility and hydraulic conductivity can be predicted by substituting the values of Q and Δs in Eq. (14.6).

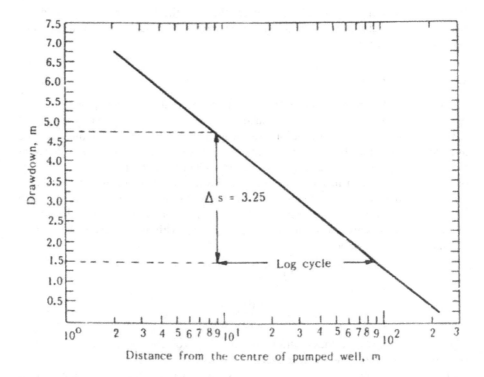

FIGURE 14.2 Distance-drawdown curve on semi-log paper for determining the transmissibility and hydraulic conductivity of aquifers.

Example 14.1

A 25 cm diameter well is pumped at a uniform rate of 3,000 L/min. The drawdown observed at 1 and 100 m distances from the center of the well are 8 and 0.4 m, respectively. Determine the hydraulic conductivity of the water-bearing strata, assuming the thickness of the saturated part of the aquifer is 25 m.

Solution:

$$Q = 3 \text{ m}^3/\text{min}$$

$$r_1 = 1 \text{ m}$$

$$r_2 = 100 \text{ m}$$

$$h_1 = H - s_1 = 25 - 8 = 17 \text{ m}$$

$$h_2 = H - s_2 = 25 - 0.4 = 24.6 \text{ m}$$

$$K = \frac{Q}{\pi \left(h_2^2 - h_1^2 \right)} \ln\left(\frac{r_2}{r_1} \right)$$

$$K = \frac{3}{\pi \left(24.6^2 - 17^2 \right)} \ln\left(\frac{100}{1} \right)$$

$$K = 0.0139 \text{ m/min}$$

Example 14.2

A 10 cm diameter well penetrates 10 m thick confined aquifer. The steady-state drawdowns were found to be 2.5 and 0.05 m at distances of 10 and 40 m, respectively, from the center of the well, when the well was operated with a constant discharge rate of 125 L/min for 12 hours. Using the Dupuit–Thiem equation, calculate the transmissibility and hydraulic conductivity of the aquifer.

Solution:

$$r_1 = 10 \text{ m}$$

$$r_2 = 40 \text{ m}$$

$$s_1 = 2.5 \text{ m}$$

$$s_2 = 0.05 \text{ m}$$

$$Q = 0.125 \text{ m}^3/\text{min}$$

Using Eq. (14.3):

$$T = \frac{Q \ln(r_2 / r_1)}{2\pi(s_1 - s_2)}$$

$$T = \frac{0.125\ln(40/10)}{2\pi(2.5-0.05)}$$

$$T = 0.0113 \text{ m}^3/\text{min}$$

Hydraulic conductivity $= T/b$

$$= \frac{0.0113}{10}$$

$$= 0.00113 \text{ m/min}$$

Example 14.3

A 30 cm diameter well completely penetrates a confined aquifer of permeability 45 m/day. The length of the strainer is 20 m. Under steady state of pumping, the drawdown at the well was found to be 3 m, and the radius of influence was 300 m.
    Calculate the discharge.

Solution:

$$r_w = 15 \text{ cm} = 0.15 \text{ m}$$

$$K = 45 \text{ m/day} = 5.208 \times 10^{-4} \text{ m/s}$$

$$b = 20 \text{ m}$$

$$s_w = 3 \text{ m}$$

$$R = 300 \text{ m}$$

Using equation:

$$Q = \frac{2\pi K b s_w}{\ln\dfrac{R}{r_w}}$$

$$Q = \frac{2 \times 3.14 \times 5.208 \times 10^{-4} \times 20 \times 3}{\ln\dfrac{300}{0.15}}$$

$$Q = 0.002583 \text{ m}^3/\text{s}$$

$$Q = 25.83 \text{ lps} = 1,550 \text{ lpm}$$

REFERENCES

Michael, A.M., Khepar, S.D. and Sondhi, S.K. 2008. *Water Wells and Pumps*. Tata McGraw-Hill Publishing Company Limited, New Delhi, India.

Raghunath, H.M. 2006. "Groundwater". *Hydrology: Principles, Analysis, Design*, Second Edition, New Age International Limited Publishers, New Delhi, India, 192–207.

Subramanya, K. 2013. "Groundwater". *Engineering Hydrology*, Shukti Mukherjee, Sandhya Chandresekhar and Sohini Mukherjee (Eds.), Fourth Edition, McGraw Hill Education Private Limited, New Delhi, India, 389–431.

15 Study of Well Losses and Well Efficiency

Objective: Study of Well Losses and Well Efficiency

15.1 WELL LOSS

In a pumping artesian well, the total drawdown at the well S_w, can be considered to be made up of three parts:

1. Head drop required to cause laminar porous media flow, called formation loss, S_{wL} (Figure 15.1).
2. Drop of piezometric head required to sustain turbulent in the region nearest to the well where the Reynolds number may be larger than unity, S_{wf}.
3. Head loss through the well screen and casing, S_{wc}.

Of these three,

$$S_{wL} \alpha\ Q \text{ and } \left(S_{wf} \text{ and } S_{wc}\right) \alpha\ Q^2$$

Thus,

$$S_w = C_1 Q + C_2 Q^2$$

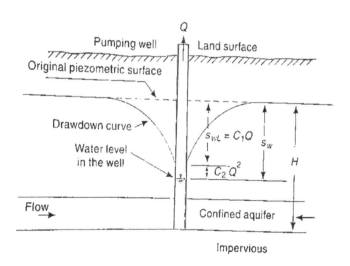

FIGURE 15.1 Definition sketch for well loss.

DOI: 10.1201/9781003319757-19

where

C_1 and C_2 are constants for the given well. While the first term C_1Q is the formation loss, the second term C_2Q is the well loss.

The magnitude of well loss has a significant impact on pump efficiency. A disproportionately high value of well loss indicates clogging of well screens and necessitates immediate action. Pump test data of drawdown for various discharges are used to calculate the coefficients C_1 and C_2 (Subramanya, 2013).

15.2   WELL EFFICIENCY

The specific capacity of a pumping well is calculated by dividing discharge by drawdown. This is a measure of the well's productivity; obviously, the higher the specific capacity, the better the well. Beginning with the Cooper–Jacob approximation of the non-equilibrium equation and incorporating the well loss,

$$S_w = \frac{2.30\,Q}{4\pi T}\log\frac{2.25\,Tt}{rw^2S} + CQ^n$$

so that specific capacity is,

$$\frac{Q}{S_w} = \frac{1}{\left(\dfrac{2.30}{4\pi T}\right)\log\left(\dfrac{2.25\,Tt}{rw^2S}\right) + CQ^{n-1}}$$

This indicates that specific capacity decreases with Q and t, as shown by the well data plotted in Figure 15.2. A well's specific capacity is frequently assumed to be constant for a given discharge. Using this approximation for $u < 0.01$ shows that the change with time is minor.

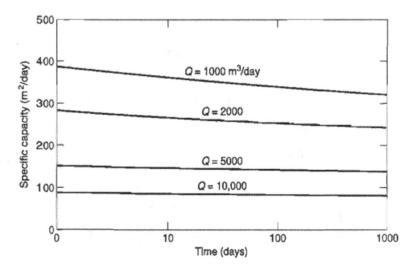

FIGURE 15.2   Variation in specific capacity of a pumping well with discharge and time.

Any significant decrease in the well's specific capacity can be attributed to either a decrease in transmissivity caused by a drop in groundwater level in an unconfined aquifer or an increase in well loss caused by clogging or deterioration of the well screen.

Specific capacity can be shown in Figure 15.3 if a pumping well is assumed to be 100% efficient. Specific capacity is plotted as a function of S, T, and a well diameter of 30 cm at the end of 1 day of pumping.

Figure 15.3 yields a theoretical specific capacity (Q/BQ) or known values of S and T in an aquifer. This computed specific capacity when compared with one measured in the field (Q/S_w) defines the approximate efficiency of the well. Thus, for a specified duration of pumping, the well efficiency (E_w) is given as percentage:

$$E_w = 100 \, \frac{Q/S_w}{Q/BQ} = 100 \, \frac{BQ}{S_w}$$

$$E_w = 100 \left(\frac{BQ}{BQ + CQ^2} \right)$$

Another way to identify an inefficient well is to take note of its initial recovery rate when pumping is stopped. When the well loss is significant, this drawdown component recovers quickly by allowing water to drain into the well from the surrounding

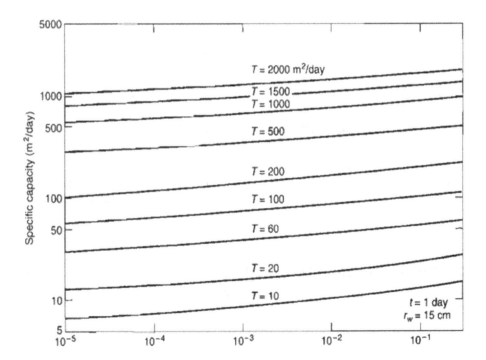

FIGURE 15.3 Graph relating specific capacity to transmissivity and storage coefficient from the Cooper–Jacob non-equilibrium equation.

aquifer. If a pump is turned off after 1 hour of pumping and 90% or more of the drawdown is recovered after 5 minutes, it can be deduced that the well is unacceptably inefficient (Subramanya, 2013).

Example 15.1

A variable rate well production test was directed with the outcomes given in Table 15.1. Calculate the coefficient of well loss and formation loss, well efficiency in the last case (Madhulika, 2018).

Solution:

For calculating the well loss, the data required are presented in Table 15.2:
Coefficient of Well Loss:

$$C = \frac{\dfrac{\Delta s_5}{\Delta Q_5} - \dfrac{\Delta s_4}{\Delta Q_4}}{Q_4 + Q_5}$$

$$C = \frac{\dfrac{1.66}{0.31} - \dfrac{2.55}{0.52}}{0.52 + 0.31} = 0.542 \text{ m}^2/\text{m}^5 = 1{,}950 \text{ sec}^2/\text{m}^5$$

Well Loss:

$$\text{Well Loss} = CQ^2 = 0.542 \times (3.27)^2 = 5.8 \text{ m}$$

TABLE 15.1
Results of Variable Rate Well Production Test

Step	1	2	3	4	5
Pumping rate (lpm)	1,590	1,980	2,440	2,960	3,270
Resulting drawdown	3.69	5.14	7.08	9.63	11.29

TABLE 15.2
Data Required for Calculating the Well Loss

Step	Q (m³/min)	ΔQ (m³/min)	S_w (m)	ΔS_w (m)
1	1.59		3.69	
2	1.98	0.39	5.14	1.45
3	2.44	0.46	7.08	1.94
4	2.96	0.52	9.63	2.55
5	3.27	0.31	11.29	1.66

Percentage of Well Loss:

$$\frac{5.8}{11.29} \times 100 = 51.4 \text{ \% of the total drawdown}$$

Formation Loss:

$$BQ = s_w - CQ^2 = 11.99 - 5.8 = 5.49$$

Percentage of Formation Loss:

$$\frac{5.49}{11.29} \times 100 = 48.6\% \text{ of the total drawdown}$$

Formation Loss Coefficient:

$$B = \frac{5.49}{3.27} = 1.67 \text{ min/m}^2$$

Well Efficiency:

$$\frac{BQ}{s_w} \times 100 = \frac{5.49}{11.29} \times 100 = 48.7\%$$

REFERENCES

Madhulika, 2018. How to Measure Well Losses. https://www.geographynotes.com/well/how-to-measure-well-losses-geography/7225

Subramanya, K. 2013. "Groundwater". *Engineering Hydrology*, Shukti Mukherjee, Sandhya Chandresekhar and Sohini Mukherjee (Eds.), Fourth Edition, McGraw Hill Education Private Limited, New Delhi, India, 389–431.

16 Computation of Interference of Wells

Objective: Computation of Interference of Wells

16.1 INTRODUCTION

Mutual interference of wells is a phenomenon in which the drawdown of interfering wells increases while their capacity decreases. If the distance between the wells is insufficient, the discharge from both wells will be reduced. The drawdown at any point in the area of influence equals the sum of the drawdown caused by each well separately (Figure 16.1).

Thus,

$$S_w = S_{w1} + S_{w2} + \ldots + S_{wn} \tag{16.1}$$

where S_w is the total drawdown at a given point and $S_{w1}, S_{w2}, \ldots, S_{wn}$ are drawdowns caused by discharge of wells 1, 2,..., n, respectively, at that point (Michael *et al.*, 2008; Todd and Mays, 2005).

16.2 INTERFERENCE OF WELLS IN CONFINED AQUIFER

For steady-state conditions, the total drawdown is given by:

$$S_w = \sum_{i=1}^{n} \frac{Q_i}{2\pi Kb} \ln \frac{Ri}{ri} \tag{16.2}$$

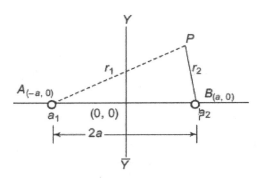

FIGURE 16.1   Two wells in a straight line.

DOI: 10.1201/9781003319757-20

where

S_w = total drawdown at a given point in the area of influence

Ri = distance from the ith well to a point at which the drawdown becomes negligible

ri = distance from the ith well to the given point

Let two wells of equal radii be located at A and B, with equal drawdown, and the spacing between them be 2a (Figure 16.1). Let r_1 and r_2 be the distances of any given point P from the centers of wells A and B, respectively. The drawdown at a point P, due to wells A and B, can be written as:

$$s_w = \frac{Q_1}{2\pi Kb} \ln \frac{R}{r_1} + \frac{Q_2}{2\pi Kb} \ln \frac{R}{r_2} \qquad (1.6.3)$$

Considering the point on the face of either well at A and B, respectively, we get:

$$s_{w1} = \frac{Q_1}{2\pi Kb} \ln \frac{R}{r_w} + \frac{Q_2}{2\pi Kb} \ln \frac{R}{2a}$$

and

$$s_{w2} = \frac{Q_1}{2\pi Kb} \ln \frac{R}{2a} + \frac{Q_2}{2\pi Kb} \ln \frac{R}{r_w}$$

Taking $S_{w1} = S_{w2} = Q_1 = Q_2$

$$H - h_w = \frac{Q_1}{2\pi Kb} \ln \frac{R^2}{r_w \times 2a}$$

$$Q_1 = Q_2 = \frac{2\pi Kb(H - h_w)}{\ln \dfrac{R^2}{r_w \times 2a}}$$

Similarly, for three wells forming an equilateral triangle spaced at distance 2a,

$$Q_1 = Q_2 = Q_3 = \frac{2\pi Kb(H - h_w)}{\ln \dfrac{R^3}{r_w \times 2a \times 2a}}$$

For three wells equally spaced at a distance of 2a, on a straight line, the discharge of the outer wells:

$$Q_1 = Q_3 = \frac{2\pi Kb(H - h_w)\ln\left(\dfrac{2a}{r_w}\right)}{\ln\left(\dfrac{R}{2a}\right)\ln\left(\dfrac{2a}{r_w}\right) + \ln\left(\dfrac{2a}{2r_w}\right)\ln\left(\dfrac{R}{r_w}\right)}$$

Whereas the discharge of the middle well:

$$Q_2 = \frac{2\pi Kb(H - h_w)\ln\left(\dfrac{2a}{2r_w}\right)}{2\ln\left(\dfrac{R}{2a}\right)\ln\left(\dfrac{2a}{r_w}\right) + \ln\left(\dfrac{2a}{2r_w}\right)\ln\left(\dfrac{R}{r_w}\right)}$$

The above equations can be applied to unconfined aquifers by replacing H by $H^2/2$ and hw by $h^2w/2$.

Thus, the discharge of each of the two wells spaced at a distance of $2a$ in an unconfined aquifer is given by:

$$Q_1 = Q_2 = \frac{\pi K\left(H^2 - h^2_w\right)}{\ln\dfrac{R^2}{r_w \times 2a}} \tag{16.4}$$

Example 16.1

Three wells, each having a diameter of 10 cm, are installed at the vertices of an equilateral triangle 10 m apart, in a confined aquifer. The radius of influence of each well is 500 m, and K is 20 m/day. The drawdown is 2 m. The thickness of the confined aquifer is 15 m. Find the discharge of each well, and the percentage decrease in discharge because of well interference.

Solution:

The discharge of each well with interference is given by:

$$Q = \frac{2\pi Kb(H - h_w)}{\ln\dfrac{R^3}{r_w \times 2a \times 2a}}$$

$$Q = \frac{2 \times \pi \times 20 \times 15 \times 2}{\ln\dfrac{500 \times 500 \times 500}{0.05 \times 10 \times 10}}$$

$$Q = 221.27 \text{ m}^3/\text{day}$$

The discharge of each well, without interference, is given by:

$$Q = \frac{2\pi Kb(H - h_w)}{\ln\dfrac{R}{r_w}}$$

$$Q = \frac{2 \times \pi \times 20 \times 15 \times 2}{\ln\dfrac{500}{0.05}}$$

$$Q = 409.03 \text{ m}^3/\text{day}$$

Therefore, reduction in discharge because of interference:

$$Q = \frac{409.03 - 221.27}{409.03} \times 100$$

$$Q = 45.9\%$$

REFERENCES

Michael, A.M., Khepar, S.D. and Sondhi, S.K. 2008. *Water Wells and Pumps*. Tata McGraw-Hill Publishing Company Limited, New Delhi, India.

Todd, K.D. and Mays, W.L. 2005. "Groundwater Movement". *Groundwater Hydrology*, Bill Zobrist, Jennifer Welter and Valerie A. Vargas (Eds.), Third Edition, John Wiley and Sons Inc., Hoboken, NJ, 86–91.

Part 5

Pumps and Their Testing

17 Study of Radial-Flow, Mixed Flow, Multistage Centrifugal Pumps, Turbine, Propeller, and Other Pumps

Objective: Study of Radial-Flow, Mixed Flow, Multistage Centrifugal Pumps, Turbine, Propeller, and Other Pumps

17.1 RADIAL-FLOW PUMP

Radial-flow pumps are centrifugal pumps in which the fluid handled exits the impeller radially. Higher centrifugal forces are generated by the flow's radial outward movement in the impeller, resulting in higher discharge pressures but typically lower volume flow rates. Liquid flows only in the radial direction through the impeller in this pump. In general, radial-flow impellers are used in all centrifugal pumps.

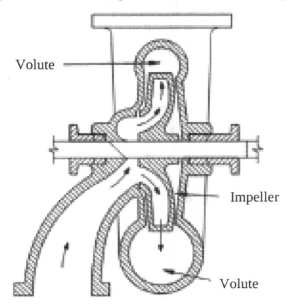

FIGURE 17.1 Radial-flow pump.

DOI: 10.1201/9781003319757-22

A radial-flow pump is used to deliver a low or medium flow rate through a larger delivery head. They are commonly used for agricultural land lift irrigation and drainage, sewage applications, cooling water for thermal and nuclear power plants, handling seawater, fresh water supply, boiler feed, mine dewatering, and other applications. Radial-flow pumps are typically supplied in horizontal configuration; however, vertical design is an option. The pump rotor is a fixed-bladed radial-flow impeller with a single or double suction (Figure 17.1) (Michael *et al.*, 2008).

17.2   MIXED FLOW PUMP

Mixed flow pump (Figure 17.2) combines aspects of both the vertical turbine pump and the propeller pump. It is suitable for medium discharge conditions with a high

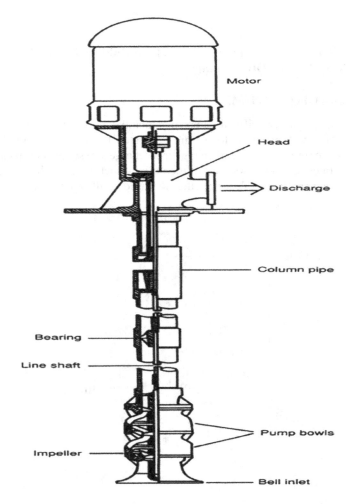

FIGURE 17.2   Two-stage mixed flow pump.

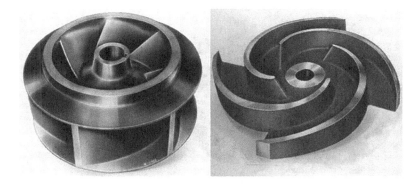

FIGURE 17.3   Impellers of mixed flow pumps.

discharge. Typically, the head ranges from 3 to 10 m. Mixed flow pumps are widely used for drainage pumping as well as pumping from canals, rivers, and streams. They are also commonly used in lift canal projects.

The specific speed of mixed flow pumps is medium. The specific speed ranges from 90 to 160 mph. Francis-type impellers operate at speeds ranging from 30 to 90 rpm.

The head of a mixed flow pump is generated in part by centrifugal force and in part by the lift of the vanes on the liquid. A single-inlet impeller is always used in this type of pump. The flow enters axially and exits both axially and radially. As a result, the pump is known as a mixed flow pump.

Mixed flow pumps can be oil or water lubricated. A mixed flow pump is built in the same way that a propeller pump is. The main distinction between the two is in the impeller construction. The impeller blades of mixed flow pumps are designed to give the water an outward thrust as well as an upward velocity. Figure 17.3 depicts the most common types of impellers used in mixed flow pumps. The diffuser located above the impeller directs the flow of water to the column and straightens it. The suction bell is circular in shape and has a flared approach for a smooth entry into the impeller (Michael *et al.*, 2008)

17.3   MULTISTAGE CENTRIFUGAL PUMP

A multistage centrifugal pump has two or more impellers mounted on a common shaft and operating in series in a single casing (Figure 17.4). The liquid flows from the discharge of the preceding impeller, or stage, to the inlet of the following impeller, causing the pressure head to rise as it passes through each stage. The use of multistage pumps in volute- and turbine-type pumps for high-head operation is standard practice.

A multistage pump's characteristics for a given type of impeller are as follows:

1. The amount of head and power required grows in direct proportion to the number of stages (impellers).

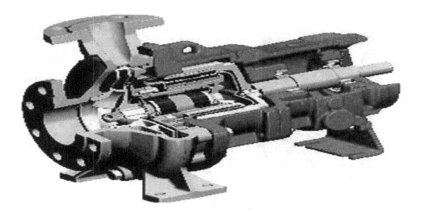

FIGURE 17.4   A two-stage horizontal centrifugal pump with top cover lifted up.

2. The discharge capacity and efficiency are nearly identical to those of a single stage of the pump working alone.

Multistage pumps are primarily used for high working heads, and the number of stages depends on the required head. Ordinary multistage centrifugal pumps typically have no more than 15 stages. A multistage pump is classified as two-stage, three-stage, four-stage, and so on based on the number of impellers mounted on a single shaft. Vertical turbine pumps with multiple stages can generate heads of up to 1,500 m. Some specially designed multistage pumps, on the other hand, can discharge up to 946 L/s and develop heads of up to 2,100 m. Remember that the head and power requirements of a multistage pump increase in direct proportion to the number of stages for a given type of impeller. The discharge and efficiency of a multistage pump, on the other hand, are nearly identical to those of a single-stage pump operating alone (Chauhan *et al.*, 1979).

17.4   DIFFUSER OR TURBINE PUMPS

The impeller in a turbine-type pump is surrounded by diffuser vanes (Figure 17.5). The diffuser vanes, like the impeller vanes, are curved and gradually enlarge to the outer end of the pump casing where the liquid enters. A large portion of the conversion of velocity energy to pressure energy occurs between the diffuser vanes in a diffuser pump. The diffuser vane casing was introduced in the design of pumps employing the water-turbine practice, where diffusion vanes are required. As a result, these pumps are frequently referred to as turbine pumps.

The choice between volute-type and turbine-type pumps is dependent on the application.

The volute-type pump is typically preferred for medium to large capacity and medium to moderately high-head applications. Turbine pumps are typically used in high-head applications. Accordingly, turbine pumps are best suited for deep tube wells due to their design advantage where the pump's diameter is small (Michael *et al.*, 2008).

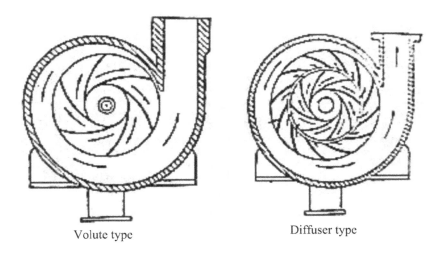

Volute type   Diffuser type

FIGURE 17.5 Diffuser or turbine-type centrifugal pump.

17.5 PROPELLER PUMP

A propeller pump is designed specifically for high-discharge, low-head pumping. It is best suited for lifting water from canals, rivers, and streams, as well as dewatering schemes. It has a high efficiency at low heads, particularly within 2 m. It can be used as portable units powered by small engines or in permanent installations powered by electric motors or engines. Portable engine-powered propeller pumps have grown in popularity in rice-growing regions throughout Southeast Asia. They have enormous potential for use in India's delta regions and canal command areas, as well as in other countries.

Propeller pumps are axial flow pumps that generate pressure head primarily through the propelling or lifting action of the propeller blades on water. The differences in the principles of operation of the pumping elements of propeller pumps and radial-flow pumps are depicted in Figure 17.6. The pressure head in radial-flow pumps (volute centrifugal pumps/turbine pumps) is generated primarily by centrifugal force. The impeller vanes direct the incoming water radially outward against the casing, where pressure is created by converting a significant portion of the velocity energy into pressure energy. A propeller pump, on the other hand, generates the majority of its head through the lifting action of the impeller, with the flow entering axially and discharging nearly axially into a guide case (Lazarkiewicz and Troskolanski, 1965).

Unfortunately, propeller pumps have not gained popularity in irrigation and drainage pumping in India. The only exception is the *petti* and *para*, which are used for drainage pumping in Kerala's backwater areas (Figure 17.7). It is a propeller pump that has been specially adapted for production by native blacksmiths and carpenters. The pump casing and discharge spout are constructed of wooden boards that are joined and reinforced with iron bands. The spout is shaped like a box. The propeller,

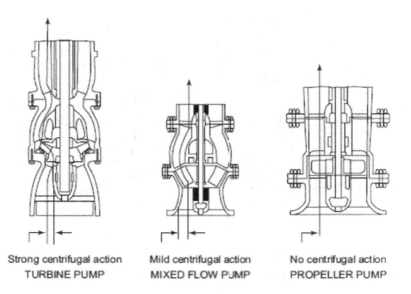

Strong centrifugal action | Mild centrifugal action | No centrifugal action
TURBINE PUMP | MIXED FLOW PUMP | PROPELLER PUMP

FIGURE 17.6 Characteristic difference in the principle of operation of turbine, mixed flow, and propeller pump.

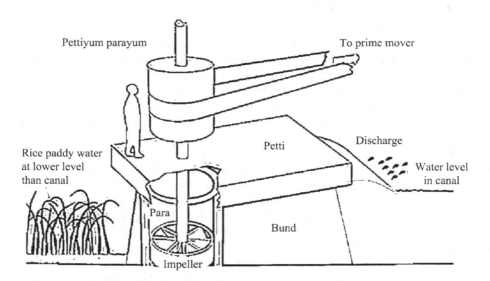

FIGURE 17.7 A locally fabricated propeller pump, called petti and para.

line shaft, and drive pulley are all made of metal. The pump, on the other hand, is inefficient and requires a lot of energy. There are only a few companies in India that manufacture factory-made propeller pumps. These pumps are designed to have specific head and discharge ranges. Prices for these pumps have been observed to be extraordinarily high and beyond the means of the average farmer. Farmers have no other option but to use the centrifugal pump, which is inefficient at low heads. As a result, the expense of irrigation or drainage is high, and there is a significant loss of energy.

17.6 JET PUMP

Jet pumps, also known as ejector pumps, are commonly used in water supply schemes to pump small to medium amounts of water at high suction lifts. In a centrifugal-jet pump combination (Figure 17.8), the centrifugal pump is mounted near the electric motor or engine, which is located on the ground surface, and provides the driving head and capacity for the jet unit, which is located in the well beneath the water surface. The jet mechanism can be installed on the ground surface (Figure 17.9) or built into the centrifugal pump casing for shallow wells up to 7.5 m deep. The centrifugal

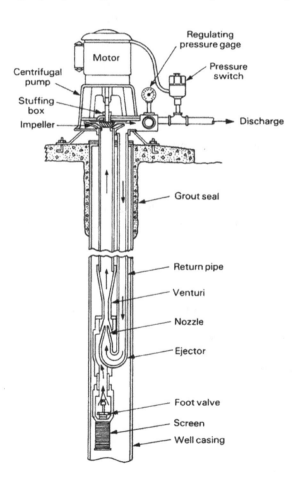

FIGURE 17.8 A centrifugal-jet pump assembly coupled to an electric motor.

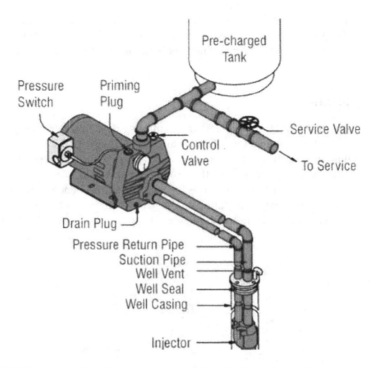

FIGURE 17.9   A centrifugal-jet pump installed in a shallow tube well.

pump-jet assembly has the advantage of having no moving parts in the well and allowing the pump-prime mover assembly to be located at a convenient point on the ground surface.

Example 17.1

A centrifugal pump has the following parameters:

- Inner diameter = 100 mm
- Outer diameter = 250 mm
- Angular speed = 1,400 rpm
- Meridional velocity = 4 m/s
- Impeller exit blade angle = 20°

Calculate the manometric head and the work required to drive the pump.

Solution:

Assuming axial inlet;
    Inlet velocity triangle:

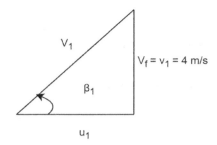

Exit velocity triangle:

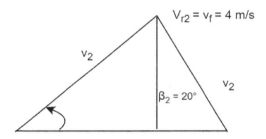

Impeller Speed:

$$u_2 = \frac{\pi d_2 N}{60}$$

$$u_2 = \frac{\pi \times 0.25 \times 1,400}{60}$$

$$u_2 = 18.33 \text{ m/s}$$

Manometric Head:

$$V_{\theta 2} = u_2 - V_{r2} \cot \beta_2$$

$$V_{\theta 2} = 18.33 - 4 \times \cot(20)$$

$$V_{\theta 2} = 7.34 \text{ m/s}$$

Pump Work:

$$gH = u_2 V_{\theta 2} - u_1 V_{\theta 1} = u_2 V_{\theta 2} \ \left(\text{since } V_{\theta 1} = 0\right)$$

$$H = \frac{u_2 V_{\theta 2}}{g}$$

$$H = \frac{18.33 \times 7.34}{9.81}$$

$$H = 13.71 \text{ m}$$

Example 17.2

Calculate the thrust that is achieved by a propeller having a diameter of 3 m with the following characteristics:

- Thrust coefficient of propeller $(C_T) = 0.048$
- Angular speed $(N) = 1,400$ rpm
- Advance velocity $= 0$
- Density $(\rho) = 1.05$ kg/m³

Solution:

Using propeller thrust force equation:

$$F_T = C_T \rho D^4 n^2$$

$$F_T = 0.048 \times 1.05 \times 3^4 \times \left(\frac{1,400}{60} \right)^2$$

$$F_T = 2223 \ N$$

Example 17.3

An axial pump has a diameter of 40 cm and has an angular speed of 1,000 rpm. If the head is 3 m and discharge coefficient of the pumps is 0.625, calculate the pump discharge.

Solution:

Using head coefficient formula:

$$C_H = \frac{\Delta h g}{D^2 n^2}$$

$$C_H = \frac{3 \times 9.81}{(0.4)^2 \times \left(\frac{1,000}{60} \right)^2}$$

$$C_H = 0.662$$

Using the formula for calculating discharge coefficient:

$$C_Q = \frac{Q}{nD^3}$$

$$Q = C_Q \times nD^3$$

$$Q = 0.625 \times \left(\frac{1,000}{60} \right) \times (0.4)^3$$

$$Q = 0.667 \ m^3/s$$

REFERENCES

Chauhan, H.S., Ram, S., Trivedi, S.K. and Sharma, H.C. 1979. *Selection, Installation and Operation of Horizontal Centrifugal Pumps for Irrigation*, Dept. of Agril. Engineering, G. B. Pant University of Agriculture and Technology, Pantnagar, 130.

Lazarkiewicz, S. and Troskolanski, A.T. 1965. *Propeller Pumps*, Pergamon Press Ltd., Oxford, UK, 501.

Michael, A.M., Khepar, S.D. and Sondhi, S.K. 2008. Water Wells and Pumps, Tata McGraw-Hill Publishing Company Limited, New Delhi, India.

Turbomachinery. Solved Questions. https://www.philadelphia.edu.jo/ academics/mebaid/ uploads/Solved_Problem_Ch14.pdf

18 Study of Installation of Centrifugal Pump Testing of Centrifugal Pump and Study of Cavitations

Objective: Study of Installation of Centrifugal Pump; Testing of Centrifugal Pump and Study of Cavitations

18.1 CENTRIFUGAL PUMP INSTALLATION

A centrifugal pump's efficiency is dependent on proper installation, operation, and maintenance. Because centrifugal irrigation pumps account for the vast majority of irrigation pumps, details pertaining to their installation under widely varying site conditions necessitate close attention. The type of installation is determined by the water source (surface water bodies/groundwater), the type of well (open well/tube well), the extent of lining in open wells, seasonal variations in static water level, and the type of prime mover used to operate the pump (electric motor/diesel engine). The Bureau of Indian Standards has established specific requirements for various pumping system components (IS: 10804–1994).

A centrifugal pump's efficiency is ensured by proper installation. Many of the pump's operational issues are the result of faulty installation. The installation of a horizontal centrifugal pump involves several steps, including its location, foundation, alignment, and piping under various site conditions (Bachus and Custodio, 2003; Chauhan *et al.*, 1979; Church and Lal, 1972).

18.2 LOCATION

The proper placement of a horizontal centrifugal pump in relation to the water level is critical. The pump should be installed as close to the water level as possible, but not submerged. As a result, the suction lift will be reduced, allowing for the use of a short and direct suction line. The pump should be installed in such a way that the total suction lift, including drawdown and friction losses, does not exceed 4.5 m, according to the Bureau of Indian Standards (IS 9694: Part 1–1987). An increase in suction lift reduces the pump's capacity and efficiency.

18.3 EFFECT OF SUCTION LIFT ON DISCHARGE AND EFFICIENCY

Figure 18.1 depicts the effect of increased suction lift on the discharge of horizontal centrifugal pumps. The discharge decreases with increasing suction lift, as shown

DOI: 10.1201/9781003319757-23

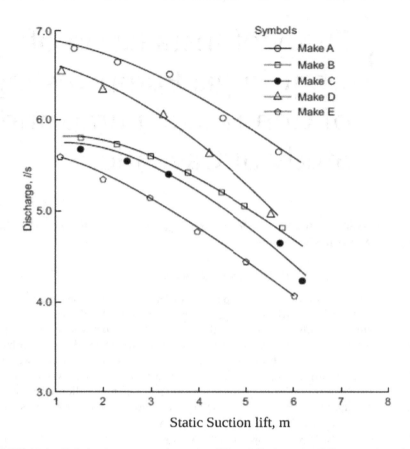

FIGURE 18.1   Relation between static suction lift and discharge in different makes of horizontal centrifugal pumps.

in Figure 18.1. The rate of decrease is slow at first, but it accelerates after a suction lift of 4.5 m. This decrease in discharge is due to a decrease in available net positive suction head (NPSH), which is a function of static suction lift. The available NPSH excess over the required NPSH decreases, which is the only source of energy for water to move up in the suction pipe.

18.4   PUMP FOUNDATION

The centrifugal pump must be built on a solid base that can sustain the unit's weight, prevent twisting or misalignment, and absorb all vibrations. The pump and prime mover are often installed on reinforced cement concrete or reinforced brick cement masonry foundation with sufficient size. Steel girders or hardwood beams, on the other hand, are sometimes utilized as pump foundations.

18.5   DESIGN CONSIDERATIONS

The following are the basic points to be considered while designing the foundation of a horizontal centrifugal pump:

1. The pump foundation should be kept independent of the foundation of the pump house.
2. The entire weight of the pumping set should be accommodated on the same foundation to avoid distortion of the machine shaft due to differential settlement.
3. The dimensions of the pump foundation are fixed so that the resultant force due to the weight of the pumping set and that of the foundation pass through the center of gravity of the base contact area.
4. The weight of the foundation and the contact area should satisfy the vibration requirements. As a thumb rule, the weight of the foundation should be at least 2.5 times the weight of the pumping set.
5. The maximum intensity of loading must be less than the allowable bearing pressure that the soil will safely carry without the risk of shear failure, irrespective of any foundation settlement that may result.
6. The surface area of the foundation should be sufficient to leave adequate space around the pump set, in order to allow grouting of foundation bolts.
7. The depth of the foundation may vary from 50 to 75 cm. However, the weight of the foundation should satisfy the criteria of 2.5 times the weight of the pump set.
8. The minimum reinforcement for the foundation should be 50 kg/m³ of concrete/brickwork.
9. The minimum diameter of the steel bars should be 12 mm. The spacing between the bars should not be more than 200 mm.
10. The diameter of the foundation bolts for the installation of a horizontal centrifugal pump varies from 12 to 20 mm, with a prime mover rating of 1.5–11 kW, respectively. The length of the bolts varies from 15 to 45 cm.
11. Usually, cement concrete 1:2:4 is used for foundation. However, for small-size pump sets, brick cement masonry with 1:4 cement mortars may be used.

18.6   DESIGN PROCEDURE

The following is the step-by-step procedure in the design of pump foundations:

1. Determine the type of soil on which the foundation is to be constructed.
2. Determine the safe bearing capacity of the soil.
3. Determine the design load which includes the following:
   a.   Dead mass of the foundation (the foundation size may be assumed to calculate the dead mass).
   b.   Total mass of pump multiplied by 3 (to take care of the dynamic forces).

4. Divide the design load by the safe bearing capacity of the soil, to determine the required base area of the foundation.
5. Determine the functional requirement of the pump, as per item (6), under design considerations.
6. The base area of the foundation selected will be either of the areas determined to satisfy the design or functional requirement, whichever is more.
7. The depth of the foundation under normal conditions may be taken to be 60 cm.
8. The actual load of the designed foundation is again calculated. The load per unit area must be less than the safe bearing capacity of the soil; otherwise, the design calculations are revised with new assumptions of the foundation size.
9. The inside diameter of the pipe sleeves for embedding the anchor bolts may be approximately two and a half times the diameter of the anchor bolt (Michael *et al.*, 2008).

18.7   ALIGNMENT

One of the most critical factors to consider when installing a pump set is shaft alignment. Though the pump and prime mover are normally aligned at the factory, the base plate can be sprung up during transit or distorted by uneven foundation bolt tightening. As a result, alignment must be checked before the pump set is turned on. Users sometimes believe that the flexible coupling will take care of misalignment. The flexible coupling does not adjust for misalignment, so keep that in mind. It is used to link the pump shaft to the prime mover as well as to transmit power to the pump. The coupling accounts for temperature variations and allows shafts to travel in opposite directions without colliding (Michael *et al.*, 2008).

18.8   INSTALLATION IN OPEN WELLS

The method for installing centrifugal pumps in open wells is determined by the prime mover employed to the depth of the pumping water level below ground level, and the well lining condition.

18.9   INSTALLATION IN SHALLOW OPEN WELLS

The overall suction lift in shallow open wells is usually restricted to 6.5 m. As a result, horizontal centrifugal pumps built on base plates and directly attached to electric motors or diesel engines can be erected at the ground surface (Figure 18.2). A pump set with belt drive is used if the prime mover speed does not meet the required pump speed for direct coupling. If the total suction lift exceeds 6 m but is within 8 m while pumping from an open well with a diesel engine, the pumping set must be installed in a shallow pit constructed adjacent to the well. For easier access, the pit is walled with brick or stone masonry and has a staircase/ladder.

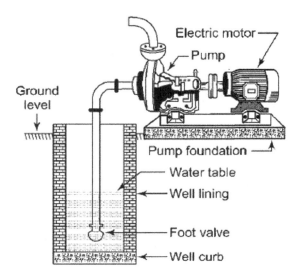

FIGURE 18.2 Typical installation of an electric motor-driven horizontal centrifugal pump in a shallow open well where the pumping water level is within 6.5 m from ground surface.

If the pumping water level is still higher, the pump must be installed in the well and the engine must be installed on the ground surface or in a shallow pit, with a flat belt drive transferring power from the engine to the pump. On the shafts of the pump and the prime mover, appropriately sized pulleys must be provided. The entire equipment should be mounted on a platform built inside the well if electric motor-driven pumps are used.

18.10 INSTALLATION IN DEEP OPEN WELLS

A horizontal centrifugal pump should be placed as close to the water table as practicable without being submerged in water. It's extremely crucial to keep electric motors out of the water.

Lined wells: The pump set, including the engine, can be put close to the static water level in lined deep open wells (Figure 18.3). The pumping station is built on a solid cantilever foundation slightly above the maximum static water level. For inspection and maintenance, a ladder is supplied. The pump's delivery pipe is held at one or more spots to prevent water hammering from damaging the pump when the motor is turned off. It may be essential to relocate the pump during the different seasons in areas where the water table fluctuates significantly between seasons. In such instances, two or more platforms in the well will be required, with the higher platform being used during periods of high water table. When the water table drops, the pumping set is lowered to the lower platform.

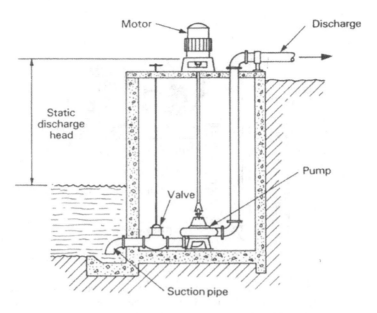

FIGURE 18.3   Motor-driven centrifugal pump in deep open well.

18.11   PARTIALLY LINED WELLS

There are times when only the upper section of the well is lined and the lower section is stable enough to keep the sides in place but not strong enough to support the pump. The pumping set is mounted on a cage frame supported by steel girders or timber beams built at the well's top in such cases (Figure 18.4). The beams are secured to the foundations, which are made of concrete or masonry. Wooden planks are commonly used to construct the platform. The cage frame can be raised or lowered to keep the pump within the maximum suction lift range. In this case, too, the basic conditions of restricting suction lifts to 6.5 m and submerging foot valves to around 45 cm below the pumping water level are met.

18.12   INSTALLATION IN TUBE WELL

The methods for installing horizontal centrifugal pumps to raise water from tube wells vary depending on the type of prime mover and the depth of the water table below ground level. In shallow tube wells with a total suction lift of less than 4.5 m, centrifugal pumps can be installed as direct coupled units at the ground surface (Figure 18.5). To aid priming, a reflux valve is installed on the suction side of the centrifugal pump. A monoblock unit or a close-coupled unit might make up the pumping

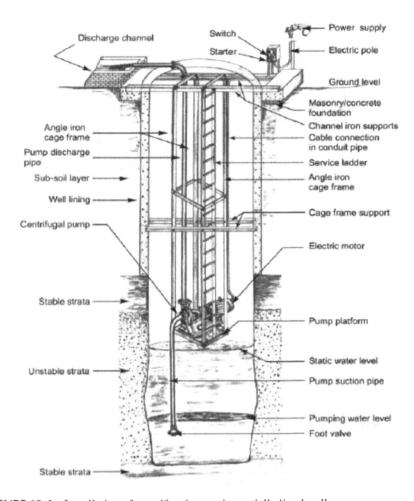

FIGURE 18.4 Installation of centrifugal pump in partially lined wells.

set. It is usual practice to employ a flat belt drive with a pulley head in engine-driven systems.

If the suction lift in a tube well is greater than 4.5 m, the pump must be positioned in a lined pit to keep it as close to the water table as possible. The pumping set, however, should always be kept above the water table to avoid the pump and prime mover being submerged. The bottom of the lined pit is left without flooring in the event of a fluctuating water table, as the water table may rise above the pit's bottom. In such cases, two or more platforms are built to locate the pumping set during various seasons.

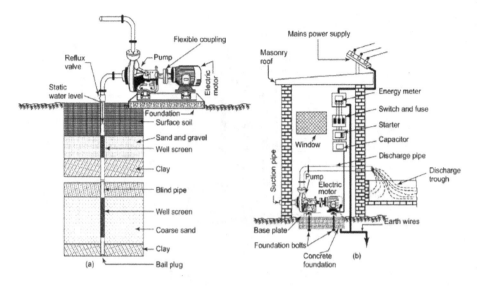

FIGURE 18.5 Installation of centrifugal pump in tube wells.

18.13 TESTING OF CENTRIFUGAL PUMP

The following tests are normally carried out for centrifugal pumps

1. Pump performance test.
2. Mechanical run test.
3. NPSHr test.

18.14 PUMP PERFORMANCE TEST

The change of pump differential head against the volumetric flow (gpm) of a liquid at an indicated rotating speed or velocity using a given amount of horsepower is plotted on a pump performance curve (brake horsepower [BHP]). On a common graph, the performance curve is actually four curves that relate to each other (Figure 18.6). These four curves are:

1. The Head-Flow Curve. It is called the H-Q Curve.
2. The Efficiency Curve.
3. The Energy Curve. It records brake horsepower.
4. The Pump's Minimum Requirement Curve. It is called Net Positive Suction Head required (NPSHr).

Pump performance testing ensures that a pump's actual performance matches the specifications provided by the manufacturer. The typical steps for conducting a pump performance test are listed below:

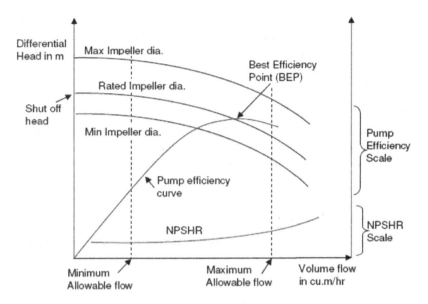

FIGURE 18.6 Performance curve.

1. Prepare the original pump curve sent by supplier.
2. Make sure that the suction strainer is clean and the suction valve is fully open.
3. Ensure that discharge valve is fully closed.
4. Start the centrifugal pump take the reading of the discharge pressure, flow rate, suction pressure, and pump Ampere. (Finish this procedure in less than 1 min so as not to damage the internal parts of the pump.)
5. Open the discharge valve slightly till the flow rate reaches the first value indicated in pump performance curve provided by pump supplier.
6. Write down the discharge pressure, flow rate, suction pressure, and pump Ampere.
7. Increase the opening of the discharge valve till you reach the next value indicated in the pump performance curve provided by pump supplier.
8. Open the discharge valve in small increments until it is fully open and take the readings of the discharge pressure, flow rate, suction pressure, and pump Ampere at each of the steps.
9. Write down the reading in the following table:

	1	2	3	4	5
Q (m³/hour)					
Suction P (bar)					
Discharge P (bar)					
Differential head (m)					
Ampere					

Examine any variations between the curve plotted using pump performance test readings (table) and the performance curve given by the supplier. If you see any deviations, attempt to figure out what's wrong and repair it before repeating the test and comparing the findings.

18.14.1 MECHANICAL RUN TEST

Until bearing temperatures have settled, the pumps will be run at their rated flow. When the bearing temperatures change by less than 2°F in 20 min, stabilization is achieved. After stabilization, a 4-hour mechanical run at the stated rated flow condition will be performed.

18.14.2 BEARING TEMPERATURE TEST

This test is usually performed before any other performance testing. Sump oil temperatures are measured during the performance test for pumps that use flood oil or ring oil lubrication. The temperature increase values are for the usual performance test, which typically lasts 1 hour or less.

The temperature of the outer bearing ring should be measured in pumps with oil mist. If no bearing temperature detectors are available, the skin temperature of the bearing housing is measured. Temperature measurements are taken in the outlet connections on the bearing housings for pumps with forced oil lubrication. Prior to the start of the performance/mechanic run test, the temperature must be measured. The temperatures reported must not exceed the temperature limit specified in the relevant codes.

The following is the maximum temperature rise over ambient temperature that can be tolerated: (Values may vary slightly depending on the pump manufacturer.)

- Pumps with flood oil or an oil ring—oil temperature of 70°F (21°C).
- Pumps with oil mist—bearing metal or bearing housing skin temperature of 70°F (21°C). The bearing oil temperature rise in pressured systems must not exceed 50°F (10°C).

18.14.3 NPSHr TEST

These are two common methods that are used to do the NPSHr test:

- Vacuum Suppression.
- Suction Valve Throttling.

18.14.3.1 Vacuum Suppression Testing

The pump is powered by a closed tank that maintains a consistent liquid level. The employment of shop air to raise pressure above atmospheric and a vacuum pump to lower pressure below atmospheric tends to change the suction circumstances.

18.14.3.2   Suction Valve Throttling

If the pump capacity exceeds that of a closed test loop and the pump is being tested on a sump, this approach is used. The pump is powered by a constant liquid level in an open sump. Depending on the characteristics of the pump, a booster pump may or may not be used to replicate suction pressure above atmospheric pressure. Throttling a valve on the pump's suction piping and also turning off the booster pump lowers the suction pressure.

The results of the vacuum method, which are unaffected by the existence of the valve, are more exact and scientific than the results of the other two procedures.

18.15   STUDY OF CAVITATIONS

If the pressure inside a pump falls below the vapor pressure, which corresponds to the temperature of the liquid, the liquid will vaporize and form vapor cavities. The vapor bubbles are carried along with the stream until they reach a higher pressure region, at which point they collapse or explode, causing tremendous shock on adjacent walls. This is referred to as cavitation.

The sudden inflow of liquid into the cavity created by collapsed vapor bubbles causes mechanical destruction, also known as 'erosion'. Aside from that, corrosion occurs as a result of a chemical reaction between the gases and the metal, resulting in additional metal destruction. There is an accompanying noise, ranging from a low rumble to loud knocks, as well as a heavy vibration of the pumping unit as a result. Power is lost as a result of the energy required to accelerate the flow of water to fill the hollow spaces. Then, cavitation is accompanied by a decrease in pump efficiency.

Cavitation will occur primarily at the vane inlet portion of the impeller, as well as on the vanes and shrouds. Gas pockets form at the point of lowest pressure, and erosion and wear due to cavitation occur further upstream, at the point of explosion. As a result, cavitation interferes with pumping. It may also cause pitting and/or excessive vibration to the pump parts. The cavitation coefficient is used to quantify cavitation. It is defined as follows:

$$\sigma = \frac{H_{sv}}{H}$$

where

H_{sv}: the net positive suction head at the critical point, m and
H: the total head, m

Cavitation can be detected by testing the pump at constant speed and capacity while varying the suction lift.

Figure 18.7 depicts the relationship between cavitation parameter σ and the percentage drop in efficiency of the selected pumps tested. As shown in Figure 18.7, the percentage drop in efficiency increases as the value of σ decreases. At higher values, there is a slight variation in the percentage drop in efficiency. When the value of σ is

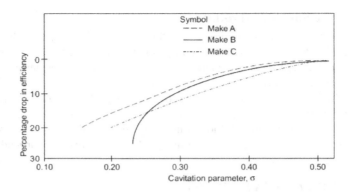

FIGURE 18.7 Relationship between cavitation parameter and drop in efficiency of pumps.

reduced below a critical limit, the drop in efficiency changes rapidly. The centrifugal pump operates successfully when a value greater than 0.3 is maintained. This critical limit of σ refers to a safe suction lift, which for most pumps is approximately 4.5–5 m.

18.15.1 NET POSITIVE SUCTION HEAD

Cavitation occurs when the hydraulic head at the pump's inlet is insufficient for the pump's operation. When water enters the pumps, the head must be sufficiently high. The pressure within the pump decreases as the velocity of flow increases. The pressure, on the other hand, should not fall below the vapor pressure of water at any point along the flow path. The required pressure head at the pump inlet is known as the NPSH or net required inlet head. The total net positive head is the sum of the NPSH and the entrance and other friction losses in the suction piping.

A pump's suction head can be less than or greater than the atmospheric pressure. When the pumping water level is lower than the pump inlet, the total NPSH equals atmospheric pressure (in meters) minus the vertical distance from the pumping water level to the pump inlet. If the pumping water level is higher than the pump inlet, the NPSH equals atmospheric pressure plus the vertical distance from the pump inlet to the pumping water level. The difference between atmospheric pressure and total NPSH determines either the maximum suction lift or the minimum submergence of the pump inlet required for proper operation. The difference between the required total NPSH and atmospheric pressure is the pump's permissible suction lift. Similarly, if the total NPSH is greater than atmospheric pressure, the difference is the minimum depth below which the pump inlet must be submerged in the well.

The NPSH is a pump characteristic. As a result, its value is independent of atmospheric pressure. The total available NPSH can be increased by lowering the pump inlet level in the well or running the pump at a higher pumping level in the well.

18.15.2 Negative Pressure

The negative pressure head, also known as suction, is the head that is lower than the atmospheric pressure head. This draws fluid into a pipe or pump chamber by creating a partial vacuum, that is, lowering the pressure below atmospheric pressure. The atmospheric pressure acting on the free surface of the water in the well forces the water up and into the pump (where the suction has been developed).

18.15.3 Maximum Suction Lift

When water flows into a pump, the maximum suction lift is limited by four factors: atmospheric pressure, vapor pressure, head loss due to friction, and the pump's NPSH. Thus,

$$H_s = H_a - H_f - e_s - \text{NPSH} - F_s$$

where

H_s: the maximum practical suction lift, m
H_a: the atmospheric pressure at the water surface, m (10.33 m at sea level)
e_s: the saturated vapor pressure of water, m
NPSH: the net positive suction head of the pump, including losses at the impeller and velocity head, m
F_s: the factor of safety, which is usually taken as 0.6 m
    A rough Ha correction for altitude is a reduction of 0.36 m for every 300 m in altitude. Suction lift may not occur even in the presence of a perfect vacuum due to other limiting factors such as vapor pressure and pipe friction. Water, like other liquids, has a proclivity to transition from the liquid to the vapor state. As a result, vapor pressure has a limiting effect on suction lift in all types of pumps.

Example 18.1

Determine the inlet and outlet dimensions and angles of a double-suction radial impeller, for an operating head of 15 m and discharge of 0.04 m³/s. The pump is to be directly connected with a motor operating at 1,450 rpm.

Solution:

Since the impeller to be designed is the double-suction type, each side of the inlet will handle a discharge of 0.02 m³/s.

1. **Specific speed**: The specific speed is computed by:

$$\eta_s = n\sqrt{\frac{Q}{H^{3/4}}}$$

$$\eta_s = 1450\sqrt{\frac{0.04/2}{15^{3/4}}}$$

$$\eta_s = 26.9$$

2. **Water horsepower**:

$$WHP = \frac{QH}{76}$$

$$WHP = \frac{0.04 \times 1000 \times 15}{76}$$

$$WHP = 7.9 \cong 8$$

The expected pump efficiency for a specific speed of 26.9 and a pump discharge of 40 L/s is 78%.
    Thus, BHP is:

$$bhp = \frac{WHP}{\eta}$$

$$bhp = \frac{8}{0.78}$$

$$bhp = 10.26$$

3. **Shaft diameter**: The shaft diameter is obtained by:

$$D_s = 3\sqrt{\frac{16T_s}{\pi f_s}}$$

Torque T_s which is based on BHP, is obtained by:

$$T_s = \frac{715\, bhp}{n}$$

$$T_s = \frac{715 \times 10.26}{1450}$$

$$T_s = 5.06 \text{ kg} - \text{m}$$

The shaft is made of mild steel, for which a safe shear stress of 3×10^6 kg/m^2 is assumed.
    Therefore,

$$D_s = 3\sqrt{\frac{16 \times 5.06}{\pi \times 3 \times 10^6}}$$

$$D_s = 0.02 \text{ m} = 2 \text{ cm}$$

The shaft diameter of 2.0 cm satisfies the requirements of torque only. With a view to satisfying the requirements of the bending moment, the shaft diameter is assumed to be 2.5 cm.

4. **Hub diameter**: The hub diameter DH is usually taken 1 cm more than the diameter of the shaft. Therefore, DH is assumed to be 3.5 cm.
5. **Impeller inlet dimension and vane angles**:
   a. **Diameter of suction flange**: The diameter of the suction flange is obtained by:

$$D_{su} = \sqrt{\frac{4Q}{V_{su}\pi}}$$

Assuming the velocity of flow, V_{su} at the suction flange as 3 m/s,

$$D_{su} = \sqrt{\frac{4 \times 0.4}{\pi \times 3}}$$

$$D_{su} = 0.13 \text{ m} = 13 \text{ cm}$$

   b. **Diameter of eye of impeller**:

$$D_0 = \sqrt{\frac{4}{\pi} \times \frac{Q'}{2C_0} + D^2H}$$

Assuming leakage losses as 2%,

$$Q' = 1.02Q$$

The velocity at the eye of the impeller is taken slightly more than the velocity of flow at the suction flange. Hence, C_0 is assumed to be 3.20 m/s.

$$D_0 = \sqrt{\frac{4}{\pi} \times \frac{1.02 \times 0.4}{2 \times 3.2} + \left(\frac{3.5}{100}\right)^2}$$

$$D_0 = 0.09 \text{ m} = 9 \text{ cm}$$

   c. **Inlet vane edge diameter**: The inlet vane edge diameter D_1 is assumed the same as the diameter of the eye of the impeller. Therefore, the value of D_1 adopted is 9.0 cm.
   d. **Passage width of the impeller at inlet**: The passage width of the impeller at the inlet is obtained from:

$$b_1 = \frac{Q}{\pi D_1 C_{m1} \in_1}$$

The contraction factor $\in_1$ is assumed to be 0.85. The radial inlet velocity C_{m1} is usually adopted slightly higher than the velocity at the eye of the impeller, C_0. Since C_0 is 3.20 m/s, the value of C_{m1} is assumed to be 3.40 m/s.

$$b_1 = \frac{0.04}{2 \times \pi \times 0.09 \times 3.4 \times 0.85}$$

$$b_1 = 0.024 \text{ m} = 2.4 \text{ cm per side}$$

e. **Inlet vane angle:**

$$\tan \beta_1 = \frac{C_{m1}}{u_1}$$

where

$$u_1 = \frac{\pi D_1 n}{60}$$

$$u_1 = \frac{\pi \times 0.09 \times 1450}{60}$$

$$u_1 = 6.82 \text{ m/s}$$

Therefore,

$$\tan \beta_1 = \frac{3.4}{6.82} = 0.5$$

$$\beta_1 = 26°33'$$

6. **Outlet vane angle and dimensions:**
   a. **Outlet diameter of impeller:** The outlet diameter of the impeller is obtained from the following relationship:

$$D_2 = \frac{84.5 \; \varnothing \; \sqrt{H}}{n}$$

Assuming the value of the head coefficient to be 1.05:

$$D_2 = \frac{84.5 \times 1.05 \times \sqrt{15}}{1,450}$$

$$D_2 = 0.24 \text{ m} = 24 \text{ cm}$$

   b. **Outlet vane angle:** The outlet vane angle β_2 is assumed larger than the inlet vane angle, to obtain a smooth and continuous passage of water in the pump casing. Therefore, the value of the vane angle β_2 adopted is 30° (usually the value of β_2 varies from 15° to 40°).
   c. **Outlet passage width:** The outlet passage width b_2 is obtained from:

$$b_2 = \frac{Q'}{\pi D_2 C_{m2} \in_2}$$

The radial velocity Cm_2 is assumed slightly less than the radial velocity at inlet Cm_1. Since Cm_1 has been assumed to be 3.40 m/s, the value of Cm_2 is assumed to be 3.25 m/s. The contraction factor is assumed to be 0.925 and the leakage loss 2%. Hence,

$$b_2 = \frac{1.2 \times 0.4}{3.25 \times 0.24 \times 3.14 \times 0.925}$$

$$b_2 = 0.018 \text{ m} = 1.8 \text{ cm}$$

The dimensions of the pump are compiled as follows:

1. Shaft diameter, $D_s = 2.0$ cm
2. Hub diameter, $DH = 3.5$ cm
3. Diameter of suction flange, $D_{su} = 13.0$ cm
4. Diameter of eye of impeller, $D_0 = 9.0$ cm
5. Velocity through the impeller eye, $C_0 = 3.20$ m/s
6. Inlet vane edge, diameter $D_1 = 9.0$ cm
7. Radial component of inlet velocity, $C_{m1} = 3.40$ m/s
8. Inlet vane angle, $\beta_1 = 26° \ 33'$
9. Outlet diameter of impeller, $D_2 = 24$ cm
10. Outlet vane angle, $\beta_2 = 30°$
11. Passage width at inlet, per side, $b_1 = 2.4$ cm
12. Passage width at outlet, $b_2 = 1.8$ cm.

Example 18.2

Determine the maximum practical suction lift for a pump having discharge of 38 L/s. The water temperature is 20°C. The total friction loss in the 10 cm diameter suction line and fittings is 1.5 m. The pump is operated at an altitude of 300 m above sea level. The NPSH of the pump, as obtained from the characteristic curves supplied by the manufacturer, is 4.7 m.

Solution:

Saturated water vapor pressure at 20°C = 0.24 m (from Standard Physical Tables).
F_s is assumed to be 0.6 m
Atmospheric pressure = 10.33 − 0.36 = 9.97 m
The maximum suction lift is given by:

$$H_s = H_a - H_f - e_s - \text{NPSH} - F_s$$

$$H_s = 9.97 - 1.5 - 0.24 - 4.7 - 0.6$$

$$H_s = 2.93 \text{ m}$$

REFERENCES

Bachus, L. and A. Custodio (Eds.). 2003. *Know and Understand Your Centrifugal Pumps*, Elsevier, Amsterdam, 264.

Chauhan, H.S., Ram, S., Trivedi, S.K. and Sharma, H.C. 1979. *Selection, Installation and Operation of Horizontal Centrifugal Pumps for Irrigation*, Dept. of Agril. Engineering, G. B. Pant University of Agriculture and Technology, Pantnagar, 130.

Church, A.H. and Lal. J. 1972. *Centrifugal Pumps and Blowers*, Metropolitan Book Co. Pvt. Ltd., New Delhi, 343.

Michael, A.M., Khepar, S.D. and Sondhi, S.K. 2008. *Water Wells and Pumps*, Tata McGraw-Hill Publishing Company Limited, New Delhi, India.

19 Study of Performance Characteristics of Hydraulic Ram

Objective: Study of Performance Characteristics of Hydraulic Ram

19.1 PRINCIPLE OF OPERATION OF HYDRAULIC RAMS

The hydraulic ram operates on the water hammer principle (Figures 19.1 and 19.2). Water hammer is a phenomenon that causes an instant increase in the pressure of water flowing in a pipe due to a sudden stoppage of its motion. The hydraulic ram's operation cycle is as follows:

By opening the gate valve on the drive pipe, the ram is started. Water flows from the supply source to the waste valve via the drive pipe. Because the waste valve is open, water is allowed to escape and flow is established along the drive pipe. Under the influence of supply head, the velocity of flow increases until the dynamic pressure on the side of waste valve is sufficient to overcome its weight. The valve then quickly closes. As a result, the supply column is retarded, resulting in a rapid increase in pressure in the valve box. This opens the delivery valve into the air vessel,

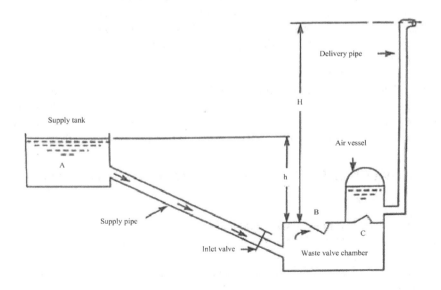

FIGURE 19.1 Principle of operation of a hydraulic ram.

DOI: 10.1201/9781003319757-24

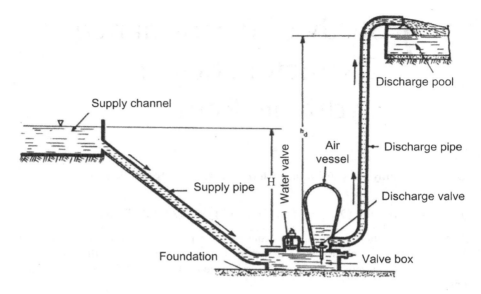

FIGURE 19.2 Components and operating principle of a hydraulic ram.

compresses the air inside, and allows water to escape via the discharge pipe. When the supply column's momentum is depleted, the compressed air and water under pressure close the delivery valve. Water flows backward in the intake chamber and drive pipe, creating a negative pressure in the valve chamber. The waste valve, which is hinged at a steep angle in the valve box, falls back due to the vacuum and its own mass, allowing water to escape. As the velocity of water in the drive pipe increases, the gate valve closes abruptly once more, and the cycle is repeated. The operation frequency ranges from 25 to 100 times per minute. It can be adjusted by adjusting a weight or other mechanism.

The supply pipe should be as long as possible in order to generate maximum impulse. If the ram is installed too close to the source of water supply, the impulse and, as a result, the delivery head will be reduced. When the pressure in the supply line falls below that of the atmosphere, an air-feeder valve in the valve box draws air into it (Ghate, 1977; Gibson, 2005; Jeffery *et al.*, 1992; Michael *et al.*, 2008).

19.2 ADVANTAGES AND LIMITATIONS OF HYDRAULIC RAMS

19.2.1 Advantages

1. The hydraulic ram is easy to build and operate.
2. It has no operating costs because it uses free energy from water and does not require any fuel or electricity.
3. It only has two moving parts: the waste valve and the delivery valve, which are both lubricated by the water itself. As a result, no separate lubrication arrangement is required.

4. It provides continuous water supply because it operates 24 hours a day, 7 days a week.
5. It performs well across a wide range of flows in the supply stream.

Hydraulic rams of smaller sizes can be made locally with simple workshop equipment.

19.2.2 LIMITATIONS

1. A minimum fall of 1 m from the stream to the ram is required for hydraulic ram installation.
2. The hydraulic ram can only lift a small portion of the flow fed to it (typically 1/20th to 1/10th of the water supplied through the drive pipe).
3. Because the ram operates 24 hours a day, a storage tank is required to store the night supply.
4. Because of the relatively high initial investment, it is better suited to ownership by small groups of farmers who can share the water among themselves, or by public agencies as part of rural water supply schemes.

19.3 EFFICIENCY OF HYDRAULIC RAMS

A hydraulic ram's efficiency can be expressed in two ways. The first expression, known as D'Aubuisson's efficiency ratio, expresses the ram's efficiency as a machine. This is the ratio of the ram's input energy to its output energy. The second expression, known as the Rankine efficiency formula, considers the ram to be a hydraulically operated pump. This takes into account the elevation difference between the water surface in the supply channel and the delivery point as the actual delivery head.

Let

q: volume of water lifted by the ram, m³/s
Q: the volume of water escaping through the waste valve, m³/s
H: the supply head (height of the source of water above waste valve), m
H_f: the head loss due to friction in supply pipe, m
h_d: the delivery head (elevation of discharge point above waste valve), m; and
h_{df}: the head loss due to friction in delivery pipe, m

i. D'Aubuisson's Efficiency Ratio

$$\frac{qh_d}{(Q+q)H}$$

A more precise expression of the efficiency of the ram, taking into account the head loss due to friction in the supply and delivery pipe is

$$\frac{q(h_d + h_{df})}{(Q+q)(H - H_f)}$$

ii. Rankine's Formula for Efficiency

$$\frac{q(h_d - H)}{QH}$$

19.3.1 Improving the Efficiency of Hydraulic Rams

The primary sources of energy loss in hydraulic rams are (i) friction and other losses in the supply pipe and valves, and (ii) velocity energy lost in the water leaving the waste valve. Both (i) and (ii) vary approximately as the square of the mean velocity of water in the supply pipe in a hydraulic ram with known supply and delivery heads. In contrast, the energy input varies directly from the mean velocity. As a result, by lowering the mean velocity of water in the supply pipe, the machine's efficiency can be increased.

19.4 INSTALLATION, OPERATION, AND MAINTENANCE OF HYDRAULIC RAMS

The selection of an appropriate site, careful planning of the various components of the system, and adherence to proper installation and maintenance procedures all contribute significantly to the economics and efficiency of a hydraulic ram (IS 10808. 1984; IS 10809. 1984; IS 11390. 1985)

19.4.1 Criteria for Selection of Site for Hydraulic Ram Installation

When deciding where to install a hydraulic ram for irrigation/water supply, keep the following factors in mind:

1. The amount of water available in the stream during cropping seasons and low flow periods.
2. The stream's available fall (supply head).
3. Elevations of water supply points in the proposed irrigation area, as well as the elevation of the water storage tank.
4. The distance between the hydraulic ram's proposed location and the ultimate delivery point of the water.
5. Maintain a safe distance from the path of potential landslides and avalanches in snow-covered areas.
6. Possibility of a stable foundation for the ram, intake tank, and drive pipe bed.
7. The total area proposed to be irrigated, the proposed population to be served, and the community's ancillary needs.
8. Cropping patterns of the irrigated area, both existing and proposed.
9. Estimated irrigation water demand.
10. Rainfall amounts and occurrence dates are expected (to estimate the shut-off periods in case the ram is used exclusively for irrigation).

19.4.2 LOCATION OF HYDRAULIC RAM

The hydraulic ram's location is chosen to have a minimum length of supply channel and pipeline. Adequate site investigations will aid in determining a location with the greatest possible vertical fall and the shortest possible supply channel length. The supply channel, which is an expensive item in the ram installation, should be as short as possible. As a general rule, the ram should be placed as close to the source of supply and the delivery point as possible while maintaining the highest vertical rainfall possible. It is sometimes preferable to sink a deep ram pit and dig out for the drive pipe and waste-flow line, resulting in a greater vertical fall and the ability to lift a larger volume of water.

A supply head is typically created in a hydraulic ram installation by digging a small counter diversion channel along the bank of a river or stream. It is common practice to build a weir and install the hydraulic ram directly below it in the case of small streams. When higher capacities are required, several hydraulic rams can be installed in parallel, with their discharge pipes connecting to a common main delivery pipe.

It is to lay the delivery pipe on the shortest possible route in order to save money on pipe costs and reduce head loss due to friction. Sharp bends in the pipeline should be avoided. Hydrants at various levels are to be installed in the delivery pipeline to irrigate fields with the least amount of pumping head. The size of the hydraulic ram and the number of rams required at a site are determined by the amount of flow that can be diverted from the stream, the magnification factor, and the amount of water required to serve the proposed cropping pattern/population. In the case of irrigation projects, where available water is limited, cropping patterns could be varied to make the best use of the available water.

The hydraulic ram's size is determined by the diameters of the supply and delivery pipes. Hydraulic rams are commercially available in India in two sizes: 10 cm × 5 cm and 30 cm × 15 cm. Larger rams could be manufactured to meet specific needs.

The manufacturer's performance data, suitability for the site, spare parts availability, and technical support are used to select a specific brand of hydraulic ram. The procedure described in the previous chapter for selecting pumping sets also applies to hydraulic rams. The size of the ram is determined by the desired output or limited by the amount of water available to power the ram. With all other factors remaining constant, the most efficient ram is chosen for installation.

The following general guidelines are followed when installing a hydraulic ram:

a. The hydraulic ram is installed level and on a solid foundation.
b. The ram is placed clearly out of the way of landslides and avalanches.
c. The ram and pipes are protected from freezing temperatures, which can be an issue at elevations above 2,400 m. When not in use, the pump and pipe should be drained of water to prevent the pipe from bursting due to freezing temperatures.
d. The hydraulic ram should be connected to the drive and delivery pipes, preferably via a union joint and a gate valve. This will make servicing the ram much easier.

e. To prevent leaks in threaded joints, wrap them in fine jute thread and then coat them with a pipe sealant compound or enamel paints.

f. Pipes should be buried below the frost line in cold climates.

19.4.3 OPERATING THE HYDRAULIC RAM

Leakage in the pipelines must be checked and repaired before the ram can be started. To start the ram, close the gate valve on the delivery pipeline and fully open the valve on the supply pipe. All of the air bubbles are allowed to escape as the water passes through the waste valve. The delivery side gate valve is then opened. It may be necessary to manually operate the waste valve for approximately 15 strokes before it begins to work automatically. A stick, pliers, or tongs can be used for this. To avoid injury from the impact of water, the valve stem is not held in place by hand. The air-feeder valve should be set so that each stroke produces a small spurt of water. If the valve is opened too far, the air chamber will fill with air and the ram will stop lifting water. If the ram is not sufficiently open, the water passing through it will absorb all of the air in the air chamber, causing the ram to pound with a metallic sound with each stroke. To avoid breakage of ram parts, this situation should be corrected as soon as possible by increasing the opening of the air-feeder valve.

The depth of waste valve opening can be adjusted using the appropriate adjustments provided in the rams (Figure 19.3). Increasing the opening of the waste valve reduces the number of strokes while increasing the amount of water used and delivered by the ram. Reduced opening increases the number of strokes while decreasing the amount of water delivered.

19.4.4 TUNING OF HYDRAULIC RAM

The number of beats of the waste valve can be changed to improve the ram's performance. This is referred to as tuning. Reduce the number of beats of the waste valve

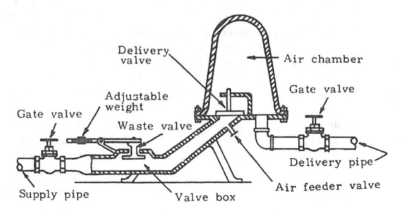

FIGURE 19.3 Sectional view of a hydraulic ram showing parts.

to improve the performance of a hydraulic ram. Higher delivery heads necessitate a lower frequency of valve cycles. The waste valve is typically set to operate at 20–40 strokes per minute in order to operate at maximum capacity. The waste valve is set to operate at 60–100 strokes per minute to regulate the ram to operate at minimum capacity under normal conditions. Rams of larger size have slower maximum and minimum strokes. Rams of larger size have slower maximum and minimum strokes. Weight is added to the valve to reduce the frequency of cycles, or the valve is adjusted by unscrewing the bolt that holds it. In the case of low- or medium-delivery heads, the weight on the valve is reduced or the bolt is tightened. Discharge measurements are taken while the impact valve is being adjusted to provide the proper setting.

19.4.5 MAINTENANCE

The hydraulic ram is a straightforward machine with only two moving parts: the waste valve and the delivery valve. It has been running for years with little maintenance. Rubber thrust pads on the waste and delivery valves may need to be replaced once or twice a year. When the machine is not in use during the winter months, the ram and pipeline must drain water to prevent pipes from bursting due to freezing water inside them. As long as the ram is in good working order, it only needs to be inspected once every 3 or 4 months to tighten the fittings, clean the ram of accumulated sediment, and check the valves for leakage and/or wear. Gate valves should be oiled twice a year, and the ram and other exposed Galvanized iron (GI) components should be painted once a year. Screens and other parts that collect sediments and other foreign objects should be inspected and serviced as needed.

Example 19.1

A hydraulic ram operates at a drive head of 3 m and a delivery head of 20 m. The flow through the drive pipe is 10 L/s and the discharge at the outlet of the outlet of the delivery pipe is 1.21 L/s. Compute the efficiency of the ram adopting (i) D'Aubuisson's ratio and (ii) Rankine's formula.

Solution

Given $H = 3$ m; $h_d = 20$ m; $Q + q = 10$ L/s;

$$Q = 10 - 1.2 = 8.8 \text{ L/s}$$

i. D'Aubuisson's efficiency ratio $= \dfrac{qh_d}{(Q+q)H}$

$$= \frac{1.2 \times 20}{(8.8 + 1.2)3} = 0.8$$

Efficiency value, $\% = 0.8 \times 100 = 80.00$

ii. Rankine's formula,

$$\text{Efficiency, } \% = \frac{q(h_d - H)}{QH} \times 100$$
$$= \frac{1.2(20 - 3)}{8.8 \times 3} \times 100$$
$$= 77.27$$

As can be seen, Rankine's formula has a lower efficiency for the ram than D'Aubuisson's ratio.

Example 19.2

Estimate the discharge of a hydraulic ram to be installed in a rural water supply system, under the following conditions:

- Flow through drive pipe = 10 L/s
- Drive head = 4.7 m
- Delivery head = 18.8 m
- Diameter of drive pipe = 10 cm
- Length of drive pipe = 34 m
- Diameter of delivery pipe = 5 cm
- Length of delivery pipe = 25 m

Accessories:

a. Gate valve in drive pipe, 1
b. Gate valve in delivery pipeline, 1
c. Number of bends, 3

Galvanized iron pipes and fittings are used in the installation. The efficiency of the ram for a lift magnification factor of 4:1 (equivalent to 18.8:4.7) as given in manufacturer's catalogs, is 62%.

Solution:

Using Darcy's formula:

$$h_f = \frac{4flv^2}{2gd}$$

The coefficient of friction is assumed to be 0.005.
Area of cross section of drive pipe:

$$a_1 = \frac{\pi}{4}d^2$$

$$a_1 = \frac{\pi}{4}\left(\frac{10}{100}\right)^2$$

$$a_1 = 0.00786 \text{ m}^2$$

Velocity of flow through drive pipe:

$$V = \frac{Q}{a_1}$$

$$V = \frac{10}{1000 \times 0.00786}$$

$$V = 1.27 \text{ m/s}$$

Head loss due to friction in drive pipe:

$$h_f = \frac{4 \times 0.005 \times 34 \times 1.27^2}{2 \times 9.81 \times \frac{10}{1000}}$$

$$h_f = 0.56 \text{ m}$$

Assuming an efficiency of 62% for the ram, as provided in the manufacturer's catalog, the approximate discharge of the ram is computed, using Rankine's efficiency formula, as follows:

$$\text{Efficiency} = \frac{q\,(h_d - H)}{QH}$$

$$0.62 = \frac{q\,(18.8 - 4.7)}{10 \times 4.7}$$

$$q = 2.07 \text{ L/s}$$

Area of cross section of delivery pipe:

$$a_2 = \frac{\pi}{4}\left(\frac{5}{100}\right)^2$$

$$a_2 = 0.002 \text{ m}^2$$

Velocity of flow in delivery pipe:

$$\frac{q}{a_2} = \frac{2.07}{1,000 \times 0.002} = 1.035 \text{ m/s}$$

Head loss due to friction in delivery pipe:

$$h_f = \frac{4 \times 0.005 \times 25 \times 1.035^2}{2 \times 9.81 \times \frac{5}{1,000}}$$

$$h_f = 0.55 \text{ m}$$

Head losses in pipeline accessories are as follows:

- 10 cm gate valve in drive pipe = 0.025 m
- 5 cm gate valve in delivery pipe = 0.016 m
- Three 5 cm bends in delivery pipe = 0.034 × 3 = 0.102 m

$$H_f = 0.56 + 0.025 = 0.585 \text{ m}$$

$$h_{df} = 0.55 + 0.016 + 0.102 = 0.668 \text{ m}$$

$$\text{Efficiency} = \frac{q\,(h_d - h_{df})}{(Q+q)(H - h_f)}$$

$$0.62 = \frac{q\,(18.8 - 0.668)}{(10+q)(4.7 - 0.585)}$$

$$q = 1.51 \text{ L/s}$$

REFERENCES

Ghate, P. 1977. Scope for irrigation through hydraulic rams in the hills of India, *Agril. Engg. Today, ISAE*, vol. 1, pp. 2–5.

Gibson, A.H. 2005. *Hydraulics and Its Applications*, Fourth Edition, Constable & Co., London, 802. http://www.nabard.org/roles/ms/mi/solar_pump.htm

IS 10808. 1984. *Code of Practice for Installation Operation and Maintenance of Hydraulic Rams*, Bureau of Indian Standards, New Delhi, 9.

IS 10809. 1984. *Specification for Hydraulic Rams*, Bureau of Indian Standards, New Delhi, 7.

IS 11390. 1985. *Test Code for Hydraulic Rams*, Bureau of Indian Standards, New Delhi.

Jeffery, T.D., Thomas, T.H., Smith, A.V., Glover, P.B. and Fountain, P.D. 1992. *Hydraulic Ram Pumps A Guide to Run Pump Water Supply Systems*, ITDG Publishing, London, 144.

Michael, A.M., Khepar, S.D. and Sondhi, S.K. 2008. *Water Wells and Pumps*, Tata McGraw-Hill Publishing Company Limited, New Delhi, India.

20 Study of Deep-Well Turbine and Submersible Pumps

Objective: Study of Deep-Well Turbine and Submersible Pumps

20.1 INTRODUCTION

Deep-well turbine and submersible pumps are vertical centrifugal pumps with diffusers that are specifically designed for pumping from tube wells. Instead of a volute, these pumps use a bowl to convert the velocity head to the pressure head. The impeller and guide vanes are housed in the pump bowl. Bowl assemblies are almost always found beneath the water's surface. As a result, deep-well turbine and submersible pumps can adapt to seasonal fluctuations in the well's water level. They are also suitable for high lifts and have a high efficiency. However, they have higher initial costs and are more difficult to install and repair than volute pumps.

A vertical turbine pump is a vertical-axis centrifugal or mixed-flow pump with stages that include a rotating impeller and a stationary bowl with guide vanes. It can be lubricated with either oil or water. The pump's main advantages are (i) it can be powered by an electric motor or an engine, and (ii) that it is less prone to wear and tear. However, the pump's initial cost is higher than that of a submersible pump with the same capacity and head. Its setup is more difficult (Johnston Pumps India, 1968; Michael, 1978; Michael *et al.*, 2008).

20.2 PRINCIPLES OPERATION OF VERTICAL TURBINE PUMPS

The turbine pump's impeller operates on a modified radial-flow centrifugal principle. Water enters the impeller near its center and is whirled by the force of the impeller. The impeller is surrounded by stationary guide vanes in the bowl. As the water exits the impeller, the gradually expanding vanes direct it upwards, partially converted to pressure by the velocity head.

The pressure head developed in a turbine pump bowl assembly, like that of all centrifugal pumps, is determined by the diameter of the impeller and the speed at which it is rotated. The relatively small diameter of the tube well limits the diameter of the bowl and the impeller inside a deep-well turbine pump. As a result, the pressure head generated by a single impeller, also known as a single-stage pump, is small. The maximum practical head per stage is approximately 10–20 m. By adding more stages, you can get more head. Because each stage is identical, the total head is calculated by multiplying the head for a specific discharge by the number of stages.

The pump's combined input energy is the sum of input energies for each stage. The relationship gives the combined efficiency of a multistage pump:

$$\eta = \frac{O(H_{s1} + H_{s2})}{102(P_{s1} + P_{s2})}$$

where

Q = the discharge capacity of pump, L/s
H_s = the head of each stage, m and
P_s = the brake horsepower for each stage, kW

A multistage deep-well turbine pump's pumping action is similar to that of a multistage centrifugal pump. Water under pressure is directed vertically upward into the center of the next higher impeller after being discharged from a lower stage. The following stage adds a comparable amount of pressure head to the water, which is then delivered to the impeller above it, and so on. The same amount of water flows from one stage to the next. It receives an additional amount of head from each stage until it finally leaves the last stage with a total head equal to the sum of the pressures it received from the individual stages and passes up the discharge column.

Turbine pumps, like volute pumps, can produce a high discharge. The capacity of the pump is affected by the design of the impeller and bowl. The capacity is determined by the area through which the flow occurs as well as the flow velocity. The velocity is determined by the impeller's peripheral speed, and the quantity is determined by the impeller's width. The greater the width of two impellers of the same diameter, the greater will be the capacity.

20.3 CONSTRUCTION OF VERTICAL TURBINE PUMPS

Vertical turbine pumps can be water or oil lubricated, with enclosed or semi-open impellers (Figure 20.1). It is made up of three major components: the pump element, the discharge column, and the discharge head.

20.3.1 PUMP ELEMENT

One or more blows or stages comprise the pump element (Figure 20.2). Each bowl assembly is made up of an impeller, a diffuser, and a bearing. The bowl assembly is still submerged beneath the water's surface. It has a screen or strainer on the bottom to keep coarse sand and gravel out of the pump. The impellers that are used are either enclosed or semi-open. Enclosed impellers allow for higher efficiencies, but they require a seal to prevent water from passing from the high-pressure side to the low-pressure side. Another benefit of the enclosed impeller is that it has a lower down thrust. To avoid excessive shaft stretch and to keep the size of the shaft and thrust bearings within reasonable limits in large pumps, the axial thrust must be reduced. Wear occurs at the suction opening on the impeller's 'eye'. The outer edges of the vanes are open on semi-open impellers. There is no need for a seal, but the edges of the impeller vanes are worn. Because of the smaller water passage, a semi-open impeller can handle suspended solids. In general, semi-open impellers

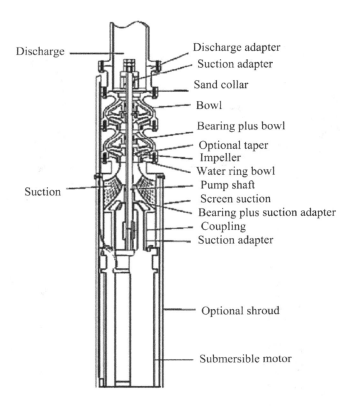

Discharge

Discharge adapter
Suction adapter
Sand collar
Bowl
Bearing plus bowl
Optional taper
Impeller
Water ring bowl
Pump shaft

Suction

Screen suction
Bearing plus suction adapter
Coupling
Suction adapter

Optional shroud

Submersible motor

FIGURE 20.1 Pump bowl assembly in water-lubricated pump with semi-open impellers.

are recommended for turbine pumps with diameters of 20 cm or less, and enclosed impellers for pumps with diameters greater than this. The impellers are made of bronze, cast iron, and porcelain enamel. Bronze, unlike cast iron, will not rust or pit. Cast iron, on the other hand, has a longer life when it is protected by porcelain enamel.

The pumping element is designed as a matched assembly, with the individual impeller design influencing the companion bowl design. Bowls are typically made of vitreous porcelain enamel close-grained and lined with cast iron. Stainless steel is commonly used for impeller shafts. Bearings are typically made of bronze.

Before starting the pump each time, columns are equipped with a pre-lubrication system consisting of a water storage tank located near the discharge head down the discharge column. This moistens the rubber bearing, preventing it from becoming overheated and damaged.

20.4 DISCHARGE HEAD OF VERTICAL TURBINE PUMP

The pump column assembly is attached to the discharge head (Figure 20.3), which is located at the ground level. The discharge head's base is bolted to the

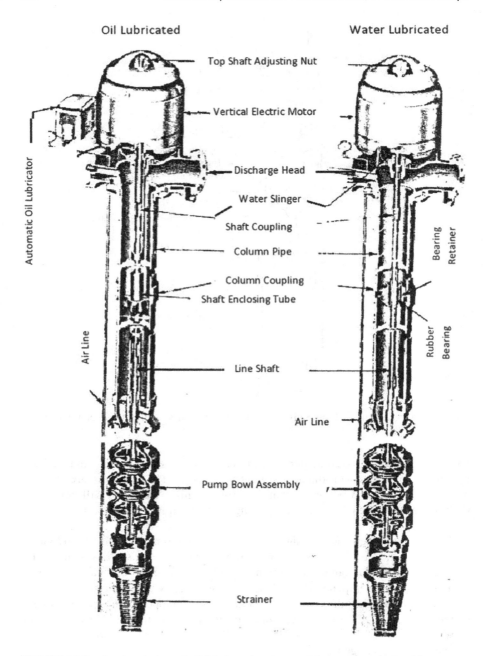

FIGURE 20.2 Sectional view of oil-lubricated and water-lubricated vertical turbine pumps showing main parts and their assembly.

pump foundations. On the same discharge head, electric motors, right angle gear drives, flat or V-belt pulleys, and suitable combinations of these drives can be used.

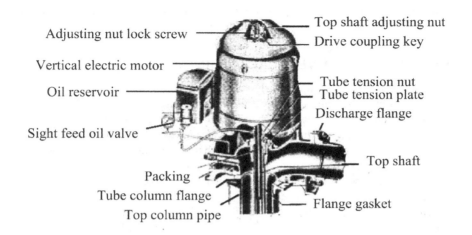

Adjusting nut lock screw — Top shaft adjusting nut / Drive coupling key

Vertical electric motor —

Oil reservoir — Tube tension nut / Tube tension plate / Discharge flange

Sight feed oil valve —

Top shaft

Packing — Tube column flange — Top column pipe — Flange gasket

FIGURE 20.3 Partially exposed view of the discharge head assembly and vertical electric motor of an oil-lubricated vertical turbine pump.

The discharge head is made of cast iron and has a simple design with large waterways. The discharge elbow has a flange to which the discharge pipe can be attached. A threaded port in the head is provided for the pre-lubrication of water-lubricated shaft bearings.

20.5 CHARACTERISTICS OF VERTICAL TURBINE PUMPS

If operated at the design speed, the efficiency curves of turbine pumps are similar to those of volute centrifugal pumps. The operating characteristics are determined by the impeller and bowl design, as well as the rotational speed. Turbine pumps cannot operate at high efficiency over the same speed range as volute pumps because high efficiency is only possible if the vanes in the bowl are parallel to the flow of water as it leaves the tip of the impeller. The direction of flow of water leaving the impeller changes when the speed of the impeller is changed. This will cause turbulence against the vane, resulting in decreased efficiency.

The impeller is designed to meet the most efficient speed, capacity, and head conditions. An impeller with narrow blades and a large ratio between the diameter of the impeller's eye and the impeller diameter would have a lower capacity but a higher rate than an impeller with a larger rate and a smaller ratio of eye diameter to impeller diameter. A high-capacity pump with a lower head would be the letter.

The pump designer can control the performance of a turbine pump by varying the design of the impeller and bowl. When these operating characteristics are determined by tests, they can be plotted on graphs to provide a clear picture of the performance of a specific impeller design. The characteristics curve serves as the foundation for selecting the appropriate type of impeller for a given application (Figure 20.3).

An inspection of the characteristics curves representing the action of a single stage of the pump under various operating conditions can be used to determine the operating characteristics of a turbine pump. If the required head exceeds the capacity of a single stage, additional stages may be added.

20.6 SELECTION OF VERTICAL TURBINE PUMPS

Today's vertical turbine pump is fairly standardized in terms of materials used and general assembly. The main difference between turbine pumps is the design of the bowl and impeller, as well as the method of lubrication. Pumps that are lubricated with oil and pumps that are lubricated with water have both been operationally successful. Oil-lubricated pumps are preferred when the pumped water contains fine sand particles. Water for drinking purposes should be oil-free, and because oil-lubricated pumps leak some oil into the water, it is preferable to use water-lubricated pumps.

In cases where the pump must frequently stand ideal for extended periods of time, an oil-lubricated pump is preferable, as water-lubricated rubber bearings can become out of shape and damaged.

Each reputable pump manufacturer has created a line of pump bowls with distinct performance characteristics. These bowls can be used singly or in series to meet any combination of head and discharge with a high degree of efficiency. One of the most important considerations in selecting a particular manufacturer's product is the main efficiency guaranteed over the specified range of pumping heads and discharge.

20.7 DATA FOR SELECTION OF A TURBINE PUMP

The selection of appropriate pump bowls and a matching column assembly will be determined by well data. The pump should be appropriate for the well's characteristics. Before purchasing a pump for permanent installation, the well should be tested. The following information is typically required to determine the type and size of pump required to fit the well's characteristics:

- Depth of well, m;
- Inside diameter of well casing, m;
- Dimension of well pit, if any, m;
- Depth to static water level, m;
- Depth to static water level at the desired capacity, m;
- Capacity of the pump, L/s;
- Drawdown-discharge curve of well;
- Data on seasonal fluctuation in water table;
- Preference for oil or water-lubricated pump design and
- **Source of power**: Electricity, voltage, phase and cycle,

  Stationary engine: Diesel/Petrol; Tractor;
  Belt-pulley size and speed.

Other details, such as the quality of water to be pumped and the condition and alignment of the well bore, are provided as needed.

Most companies that sell vertical turbine pumps have their own data from which the pump parts are chosen and matched to meet a specific condition. Improper pump selection can cause both hydraulic and mechanical problems.

20.8 SUBMERSIBLE PUMPS

A submersible pump is a vertical turbine pump linked to a small-diameter submersible electric motor. The motor is mounted right beneath the pump's intake. The pump and motor work while being completely submerged. The large vertical shaft in the column pipe is eliminated with this arrangement. The vertical turbine pump's performance characteristics are comparable to those of the submersible pump. The direct coupling of the motor and its good cooling by submersion in water improve the pump's efficiency.

The submersible pump's main advantage is that it can be utilized in very deep tube wells where a long shaft would be impractical. Deviations in the well's vertical alignment have less of an impact on these pumps. Because there are no aboveground operating elements on the submersible pump, it can be employed in areas where flooding is a concern. It can also be used in places where an aboveground dwelling would be impractical, such as public grounds. A submersible pump can be put in a well as tiny as 10 cm in diameter. Its fundamental disadvantage is that it cannot be powered by an engine, limiting its application to areas with access to electric power.

20.8.1 Constructional Details

The different parts of submersible pump are pump bowl and electric motor assembly, a discharge column, a head assembly, and a waterproof cable to conduct current to the submerged motor.

20.8.1.1 Submersible Electric Motor

The diameter of the submersible electric motor matches that of the pump bowl. It's a lot longer than a regular motor with the same horsepower (Figure 20.4). It's a squirrel-cage induction motor, and it can be dry or wet. The dry motor is housed in a steel box that is filled with high-electric-strength light oil. At the point where the drive shaft passes through the housing of the impeller, a mercury seal installed precisely above the motor armature prevents oil leakage or water entry.

Wet motors are ones in which well water has entry to the inside of the motor and the rotor and bearings are really submerged. The stator windings are fully sealed off from the rotor in this type of motor by a thin stainless steel inner liner. To prevent abrasive particles from entering the motor, a filter around the shaft is required. The wet-type motor should be filled with water during installation to ensure that the bearings are properly lubricated when the motor is turned on for the first time.

The stator windings are continuous throughout the motor's length. The rotors are manufactured in parts on a continuous shaft, with bearings in between to guide the shaft and keep it aligned. The electric line that connects the motor to the ground-level switchboard is waterproofed and positioned outside the discharge pipe.

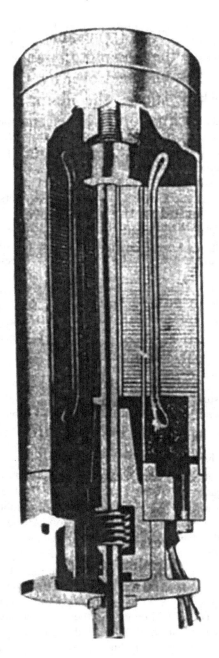

FIGURE 20.4 A submersible pump electric motor.

20.8.1.2 Pump Elements

Multistage centrifugal pumps with radial-flow or mixed-flow impellers are known as submersible pumps.

Pumps with radial-flow impellers are utilized for high total heads and low capacities, while pumps with mixed-flow impellers are used for medium capacities and medium heads. The impellers are dynamically balanced, and all of the pump's bearings are lubricated with water and shielded from sand and other suspended particles. A perforated strainer protects the pump suction casing between the pump and the motor from any suspended debris in the water. Figure 20.5 shows an exploded view of the pump element.

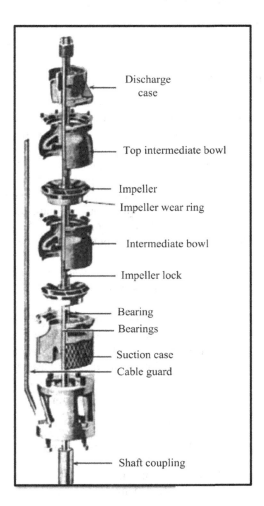

FIGURE 20.5 Exploded view of pump element.

20.8.1.3   Discharge Column and Head Assembly

The discharge column is made up of a discharge pipeline that is linked to the pump element. The discharge pipe is supported at ground level by a supporting clamp that is supported by a solid foundation. To indicate the total operating head, a pressure gauge is sometimes attached to the bend in the delivery pipe.

A sluice valve is usually installed in the delivery pipe to control the flow. The power cable is connected to switchgear that is mounted on a switchboard. The switchgear is typically oil-filled and consists of three-phase contactors and an automatic star-delta starter.

20.9   INSTALLATION OF SUBMERSIBLE PUMP

The installation of submersible pump consists of initial preparations, installation, and testing.

20.9.1   Initial Preparations

Prior to introducing the pumping set, it ought to be guaranteed that the well is completely evolved and the release water is liberated from sand. The water level and drawdown ought to be learned. The well ought to be tried for its arrangement. A slight tendency of the bore opening is irrelevant; however, a slanted segment can make the establishment troublesome or even impossible.

Unload and inspect the pumping set and accessories for any harm or deficiency. For the most part, packing directions are provided by the makers. Inspect the link for harm especially close to the pump. The link should be painstakingly taken care of consistently, and ought not to be pulled, curved, hauled over sharp stones, or run over by wheeled traffic. Most importantly, the link ought to never be permitted to convey the heaviness of the unit.

Test the engine for congruity and insulation resistance through a Megger, to guarantee that it has not been harmed on the way. The insulation resistance ought to surpass $10\,M\Omega$.

Submersible motors have water-greased up plain bearings and must never dry up. In any event, testing the motor for a brief period out of water should be kept away from it, as it might make harm the course. Prior to introducing it in the well, the motor should be filled in an upstanding position, with clean water liberated from sand and corrosive.

20.9.2   Installation Procedure

The installation procedure changes as per the size of the submersible pump unit. Generally, the steps included are as follows.

The initial step is to clasp, at a helpful level, one of the sets of supporting clips (provided with the pump), to the main length of the riser pipe. The riser pipe is then associated with the release branch or check valve on the pump. Through the tackle, the pump is then leisurely brought down into the well until its weight is taken by the supporting braces laying on two equal gentle steel channels or wooden light emissions aspects, put ready across the highest point of the well to go about as conveyors.

The tackle of the pulley block is then disconnected. One more length of line is given to the second sets of supporting clips. The line is brought down leisurely, utilizing the chain pulley block associated with the main length of line, and joined to the tackle. The main sets of supporting clasps can then be eliminated, moving the weight of the assembly onto the tackle, through which the pump and piping assembly is gradually brought down until the weight is transferred to the second pair of clamps. The activity is rehashed for every length of the section pipe until the ideal depth is reached.

The strung piped segments should be firmly screwed and, if vital, a locking gadget must be fitted so that the pump unit can't fall into the well through one of the lines getting unscrewed during pump activity. It should be recollected that, while turning on and off habitually, a curving moment is applied on the segment pipe.

During installation, a couple of clips should constantly be immovably connected to the riser pipe so that, in case of failure of the lifting gear, the pump can't fall into the well.

The depth at which the pump is introduced ought to be to such an extent that the top spine of the pump is something like 50 cm underneath the most minimal degree of water in the well. This level is reached after the pump has been running for a delayed period with the valve completely open.

20.9.3 OPERATION OF SUBMERSIBLE PUMP

Prior to working the pump, the direction of rotation is analyzed. To find out the right direction of rotation, let the engine run in the two directions with the stop valve shut. The direction of rotation is changed by trading two of the stages. The strain check will show various readings for the two directions. The direction which gives the higher tension is the right one. While running openly, the right direction of rotation can likewise be decided from how much water is pumped.

REFERENCES

Johnston Pumps India. 1968. *Johnston Vertical Turbine Pumps: Installation, Operation and Maintenance Manual*, Johnston Pumps India Pvt. Ltd., Kolkata, India.

Michael, A. M. 1978. *Irrigation Theory and Practice*, Vikas Publishing House Pvt. Ltd., New Delhi, India.

Michael, A.M., Khepar, S.D. and Sondhi, S.K. 2008. Water Wells and Pumps, Tata McGraw-Hill Publishing Company Limited, New Delhi, India.

21 Analysis of Pumping-Test Data

Objective: Analysis of Pumping-Test Data

21.1 INTRODUCTION

Different types of pumping tests are available which provide varying types of pumping-test data for confined, unconfined, and leaky aquifer systems. Depending on the type of pumping-test data and the type of aquifer in which the test is conducted, a wide range of methods are available for analyzing pumping-test data in order to determine aquifer parameters. Table 21.1 summarizes commonly used methods for analyzing pumping-test data obtained from confined, unconfined, and leaky confined aquifers (Michael *et al.*, 2008; Kruseman and de Ridder, 1994).

TABLE 21.1
Commonly Used Methods for Pumping-Test Data Analysis

No.	Type of Aquifer	Type of Pumping-Test Data	Name of Methods
1	Confined Aquifer	i. Time-Drawdown data	• Theis Type-Curve Method • Cooper–Jacob Straight-Line Method
		ii. Unsteady Distance-Drawdown data	• Cooper–Jacob Straight-Line Method
		iii. Quasi-Steady/Steady Distance-Drawdown data	• Thiem Method • Graphical Method
		iv. Recovery data: • Time-Residual Drawdown data • Time-Recovery data	• Residual Drawdown-Time Ratio Method • Cooper–Jacob Straight-Line Method
2	Unconfined Aquifer without Delayed Yield	i. Time-Drawdown data	• Theis Type-Curve Method • Cooper–Jacob Straight-Line Method
		ii. Unsteady Distance-Drawdown data	• Cooper–Jacob Straight-Line Method
		iii. Quasi-Steady/Steady Distance-Drawdown data	• Thiem Method • Graphical Method
		iv. Recovery data: • Time-Residual Drawdown data • Time-Recovery data	• Residual Drawdown-Time Ratio Method • Cooper–Jacob Straight-Line Method

(Continued)

DOI: 10.1201/9781003319757-26

TABLE 21.1 (*Continued*)
Commonly Used Methods for Pumping-Test Data Analysis

No.	Type of Aquifer	Type of Pumping-Test Data	Name of Methods
3	Unconfined Aquifer with Delayed Yield	i. Time-Drawdown data	• Type-Curve Method • Neuman Straight-Line Method
		ii. Quasi-Steady/Steady Distance-Drawdown data	• Thiem Method • Graphical Method
4	Leaky Confined Aquifer without Storage in Aquitards	i. Time-Drawdown data	• Walton Type-Curve Method • Hantush Inflection-Point Method
		ii. Quasi-Steady/Steady Distance-Drawdown data	• Type-Curve Method
5	Leaky Confined Aquifer with Storage in Aquitards	i. Time-Drawdown data	• Hantush Type-Curve Method
		ii. Quasi-Steady/Steady Distance-Drawdown data	• Type-Curve Method

21.2 TYPES OF PUMPING TESTS

Normal kinds of pumping tests that you might perform incorporate the accompanying:

21.2.1 CONSTANT RATE TEST

Steady rate tests keep up with pumping at the control well at a consistent rate. This is the most usually utilized pumping-test technique for acquiring evaluations of aquifer properties.

21.2.2 STEP-DRAWDOWN TESTS

These continue through a succession of steady rate ventures at the control well to decide well execution qualities like well misfortune and well proficiency.

21.2.3 RECUPERATION TESTS

These use water-level (residual drawdown) estimations after the end of pumping. Albeit frequently deciphered independently, a recuperation test is a necessary piece of any pumping test.

21.3 DESIGNING A PUMPING TEST

The accompanying surveillance work and arranging ought to factor into the plan of a siphoning test (Analysis of pumping test data, Anonymous, 2007):

- Best season for pumping test
- Geography of appropriate aquifer and aquitard units
- Normal specialists of groundwater vacillation, for example, barometric strain changes, earth tides, and flowing varieties which might influence water levels in perception wells during the pumping test
- Off-site groundwater clients which might impact water levels during the pumping test
- Area and development subtleties of new and existing wells
- Depth setting and sort of pump in charge well
- Pumping length and rate
- Pumping rate estimation and control
- Water-level estimation and recurrence
- Procedure for removal of pumped water
- Assortment of water quality examples during pumping
- Assessments of aquifer properties from past work or writing information
- Area and direction of streams, shortcomings, lithologic contacts, and other potential spring limits
- Potential for salt water interruption in seaside regions
- Field hardware
- Forecast of pumping-test reaction
- Effect of pumping test on neighboring water clients

REFERENCES

Anonymous. 2007. Analysis of Pumping Test Data. http://ecoursesonline.iasri.res.in/mod/page/view.php?id=1836

Kruseman, G.P. and de Ridder, N.A. 1994. *Analysis and Evaluation of Pumping Test Data.* Second Edition, ILRI Publication 47, International Institute for Land Reclamation and Improvement (ILRI), Wageningen, the Netherlands.

Michael, A.M., Khepar, S.D. and Sondhi, S.K. 2008. *Water Well and Pump Engineering.* Second Edition, Tata McGraw Hill Education Pvt. Ltd., New Delhi, India.

Part 6

Groundwater Modeling

22 Groundwater Computer Simulation Models

Objective: Groundwater Computer Simulation Models

22.1 OVERVIEW OF GROUNDWATER MODELING

Groundwater models are computer simulations that depict the processes that take place in the natural groundwater surroundings in a simplified manner.

In light of specific working hypotheses, groundwater models use numerical circumstances to represent the groundwater flow and transport processes. The direction of flow, the aquifer's geology, the heterogeneity or anisotropy of the sediments or bedrock within the aquifer, the foreign substance transport mechanisms, and the substance reactions are typically included in these hypotheses. A model should be viewed as an approximation and not a meticulous replication of field conditions due to the working on assumptions implanted in the numerical conditions and the multiple weaknesses in the upsides of information required by the model. Despite this, groundwater hydrologists may employ approximations and models as a useful evaluation tool for a variety of applications. The use of existing groundwater models includes water balance (in terms of water amount), learning about the quantitative components of the saturated zone, simulating water stream and material movement in the soaked zone, including stream groundwater relationships, assessing the impact of changes to the groundwater system on the climate, setting up/advancing checking organizations, and setting up groundwater assurance zones.

It would be impossible to assess all of the natural processes that affect a hydrogeologic design without groundwater models due to the complexity in the physical processes that occur in the hydro-geologic environment, the spatial distribution of properties and boundaries, and the temporal nature of the flow system.

A groundwater model can be used to examine or explore "what if" questions about how the hydro-geologic system will react to future design changes once it has been built and calibrated. For instance, you can incorporate a groundwater design into the model to assess its effects on hydraulic heads and flow before it is built in the field, saving time and money. This design might include pumping wells and injection wells, infiltration galleries, grout cut-off walls, seepage drains, and other features (Kumar, 2001; Groundwater modeling, Anonymous, 2020).

22.2 REPRESENTATION OF GROUNDWATER SYSTEM IN A COMPUTER MODEL

All of this complexity can be incorporated into a groundwater model, which can be used to evaluate various alternatives and future situations. It is vital to translate the

DOI: 10.1201/9781003319757-28

physical environment into the modeling application while constructing a groundwater model. Hydro-geologic factors such as conductivity and storativity are derived from geology. Boundary conditions in a model are hydrologic boundaries that have an impact on the groundwater flow system, such as recharge zones, rivers, lakes, and wells. Field observations, such as groundwater levels, fluxes, or contaminant concentrations, are used to calibrate the model in the physical world, allowing it to match what is seen in the real world as precisely as possible.

22.3 USES OF GROUNDWATER MODELS

- Well systems and water resources can be evaluated using groundwater models (yield, drawdown, interference, etc.).
- Describe the capture zones and wellhead protection zones (WHPAs).
- Examine the effects of mine dewatering on the ecosystem.
- Examine the quantity and quality of water at mine sites.
- Examine the consequences of polluted places (industrial sites, leaky USTs, spills).
- Examine the consequences of landfill contamination.
- Create/improve remedial solutions for contaminated groundwater.
- In coastal aquifers, assess and limit the effects of saltwater intrusion.
- Construction and Excavation Dewatering.
- Calculate the transport of heat in the subsurface and the potential for geothermal energy.
- Septic System, Sewage Effluent, and Agricultural Practices Nitrate Fate and Transport Modeling.

22.4 GROUNDWATER FLOW EQUATION

Groundwater modeling starts with a calculated comprehension of the physical issue. The subsequent stage in modeling is making an interpretation of the actual framework into numerical terms. As a general rule, the outcomes are the natural groundwater flow equations and transport conditions. The leading flow equation for three-dimensional saturated flow in saturated porous permeable medium is:

$$\frac{\partial}{\partial x}\left(K_{xx}\frac{\partial h}{\partial x}\right)+\frac{\partial}{\partial y}\left(K_{yy}\frac{\partial h}{\partial y}\right)+\frac{\partial}{\partial z}\left(K_{zz}\frac{\partial h}{\partial z}\right)-Q=S_s\frac{\partial h}{\partial t}$$

where

K_{xx}, K_{yy}, K_{zz} : Hydraulic conductivity along the x, y, z axes which are assumed to be parallel to the major axis of hydraulic conductivity

h : Piezometric head

Q : Volumetric flux per unit volume representing source or sink terms

S_s : Specific storage coefficient

The transport of solutes in the saturated zone is represented by the advection–dispersion equation which for a permeable medium with uniform porosity circulation is figured out as follows:

$$\frac{\partial c}{\partial t} = -\frac{\partial}{\partial x_i}(cv_i) + \frac{\partial}{\partial x_i}\left(Dij\frac{\partial c}{\partial x_j}\right) + R_c$$

where

> c : Concentration of solute
> R_c : Source or sink
> D_{ij} : Dispersion coefficient tensor
> v_i : Velocity tensor

Before a modeling problem can be created, it is essential to understand these equations and the boundary and beginning conditions that are associated to them. Groundwater flow, solute transport, and heat transfer are considered to be fundamental cycles. The majority of groundwater modeling investigations are conducted using either deterministic models, which reflect the probabilistic notion of a groundwater framework, or stochastic models, which reflect the exact description of situations and logical conclusions or input-response connections.

Groundwater framework controlling equations are often resolved analytically or mathematically. Scientific models include logical field conditions that are consistently present in reality. By using mathematical approximations of the governing partial differential equations, a discrete arrangement is obtained in both reality spaces in mathematical models. In groundwater models, several mathematical arranging techniques are used. The finite-difference method, finite-element method, and logical element method are three of the more complex groundwater modeling techniques. Regarding accessibility, prices, convenience of use, applicability, and user-required information, each strategy has its own advantages and disadvantages.

22.5   GROUNDWATER FLOW MODELS

The following provides notable highlights of the most popular groundwater models. The U.S. Topographical Survey originally developed the three-dimensional MODFLOW mathematical groundwater flow model, which is the most widely used (McDonald and Harbaugh, 1988). For the saturated zone, it uses a block-focused, limited contrast plot. The benefits of MODFLOW include a range of information readiness facilities, straightforward information exchange in a standardized format, a broader overall experience with persistent events, accessibility to the source code, and a somewhat low cost. In the event of transitory problems, MODFLOW cannot be used because the transition at the groundwater table depends on the determined head and the capability isn't known in advance. However, surface overflow and unsaturated streams are eliminated.

22.5.1 FEFLOW (Finite-Element Subsurface Flow System)

FEFLOW is a finite-element software program used to simulate subsurface mass (contaminant salinity) and heat transport in both 3D and 2D fluid density coupled flow. It is competent of computing:

- Problems in saturated–unsaturated zones;
- Groundwater systems with and without open surfaces (phreatic aquifers, perched water tables, and moving meshes);
- Thermohaline fluxes, which are both salinity- and temperature-dependent transport processes;
- Situations with complex parametric and geometric structures.

The software is entirely interactive and graphics based. Integration of pre-, main-, and post-processing is used. GIS (Geographic Information System) has both a programming interface and a data interface. Large issues can be solved thanks to the numerical characteristics that have been implemented. Techniques for adaptation are used.

22.5.2 HST3D (3D Heat and Solute-Transport Model)

The Argus Open Numerical Environments (Argus ONE) modeling platform includes a robust user-friendly interface for HST3D. The user can enter all spatial data into HST3D, run HST3D graphically, and view the outcomes. Within a single complete graphical user interface, Argus ONE unifies CAD, GIS, databases, conceptual modeling, geo-statistics, automatic grid and mesh generation, and scientific visualization (GUI). The three-dimensional simulation of groundwater flow and related heat and solute transport is done by the Heat and Solute-Transport Model HST3D. The examination of issues such subsurface waste injection, landfill leaching, saltwater intrusion, freshwater recharge and recovery, radioactive waste disposal, water geothermal systems, and subsurface energy storage may be done using the HST3D model. HST3D needs the Argus ONE GIS and Grid Modules to function.

22.5.3 MODFLOW (Three-Dimensional Finite-Difference Groundwater Flow Model)

The USGS Modular Three-Dimensional Groundwater Flow Model is also known as MODFLOW. MODFLOW has evolved into the accepted global standard groundwater flow model due to its capacity to simulate a wide range of systems, its vast publicly available documentation, and its stringent USGS peer review. Systems for water delivery, containment cleanup, and mine dewatering are all simulated using MODFLOW. The accepted standard model is MODFLOW when used correctly. The main goals in creating MODFLOW were to create a program that can be easily adjusted, is straightforward to use and maintain, can be used to run huge problems on a range of systems with little modification, and can handle large data sets. The MODFLOW article demonstrates in great detail the physical and mathematical principles upon which the model is built, as well as how those principles were incorporated into the computer program's modular design. A Main Program and a number of modules—highly independent subroutines—make up MODFLOW's modular structure. Packages are used to group the modules. Each package focuses on a particular aspect of the hydrologic system that needs to be modeled, such as the flow into or out of drains or rivers, or on a particular approach to solving the linear equations that describe the flow system, such as the Strongly Implicit Procedure or Preconditioned Conjugate Gradient. The modular structure of MODFLOW enables the user to individually study particular model hydrologic properties. The ability to add new modules or packages to the application without altering the existing ones makes it easier to develop new capabilities. The MODFLOW input/output system was created with maximum flexibility in mind. A block-centered finite-difference method is used in MODFLOW to simulate groundwater flow within the aquifer. Simulated layers can be constrained, unconstrained, or a combination of the two. It is also possible to mimic flows caused by external stresses such flow to wells, areal recharge, evapotranspiration, flow to drains, and flow across riverbeds.

22.5.4   MT3D (A Modular 3D Solute-Transport Model)

The MT3D numerical model is a thorough three-dimensional model for simulating solute transport in intricate hydro-geologic situations. The modular nature of MT3D enables the standalone or collaborative simulation of transport processes. Advection in intricate steady-state and transient flow fields, anisotropic dispersion, first-order decay and production reactions, and both linear and nonlinear sorption may all be modeled using MT3D. Additionally, it can manage daughter products, monad reactions, and bioplume-type reactions. As a result, MT3D is able to mimic or evaluate natural attenuation within a contaminated plume as well as perform multi-species responses. In order to solve advectively-dominated transport issues, MT3D is connected to the USGS MODFLOW groundwater flow simulator. This eliminates the need to build complex models specifically for solute transport.

22.5.5   SEAWAT (Three-Dimensional Variable-Density Groundwater Flow)

To simulate three-dimensional, variable-density, transient groundwater flow in porous media, the SEAWAT program was created. In order to solve the linked flow and solute-transport equations, MODFLOW and MT3DMS were combined to create the source code for SEAWAT. Due to the modular design of the SEAWAT code, new features can be introduced with just small changes to the core program. Standard MODFLOW and MT3DMS data files can be read and written by SEAWAT, while some SEAWAT simulations may require additional input. As a result, numerous pre- and post-processors already in use can be utilized to generate input data sets and analyze simulation results. Users accustomed to MODFLOW and MT3DMS should have no trouble using SEAWAT to solve issues involving variable-density groundwater flow.

22.5.6   SUTRA (2D Saturated/Unsaturated Transport Model)

A full saltwater intrusion and energy transfer model, SUTRA is a 2D groundwater saturated–unsaturated transport model. In a subterranean setting, SUTRA simulates fluid movement and delivery of either energy or dissolved chemicals. The governing equations for the two interdependent processes that are simulated by SUTRA are approximated using a two-dimensional hybrid finite-element and integrated finite-difference method:

- Transport of thermal energy in the groundwater and solid matrix of the aquifer; fluid density dependent saturated or unsaturated groundwater flow; or
- Transport of a solute in the groundwater, in which the solute may be subject to equilibrium adsorption on the porous matrix and both first-order and zero-order production or decay.

Recently, SUTRA was made available in 3D.

22.5.7   SWIMv1/SWIMv2 (Soil Water Infiltration and Movement Model – Simulate Soil Water Balances)

A software program called SWIMv1 (Soil Water Infiltration and Movement model version 1) simulates water infiltration and movement in soils. With the help of numerical solutions to the fundamental soil water flow equations, the user of the menu-driven SWIMv1 suite of three programs can model soil water balances. SWIMv1 permits both the input of water to the system through precipitation and its removal through runoff, drainage, evaporation from the soil surface, and transpiration by vegetation, just like in the real world. Researchers and consultants can evaluate the implications of procedures like tree cutting, strip mining, and irrigation management by using SWIMv1, which aids in their understanding of the soil water balance. Scientists and consultants working in land planning and management can benefit from SWIMv1. For instance, SWIMv1 can be used to identify salinity or surface runoff issues that may arise from a change in the soil water balance brought on by the removal of the trees in a development that is under consideration. A mechanistically based model called SWIMv2 (Soil Water Infiltration and Movement Model Version 2) was created to address problems with the balance of soil water and solutes that are related to both production and the environmental effects of production. In order to solve the convection-dispersion equation for solute transport as well as Richards' equation for water flow, SWIMv2 uses quick, numerically effective algorithms. The model deals with a one-dimensional vertical soil profile that may be inhomogeneous vertically but is considered to be uniformly distributed horizontally. It has the ability to model runoff, deep drainage, leaching, plant absorption and transpiration, infiltration, redistribution, solute transport, and redistribution of solutes.

22.5.8 Visual HELP (Modeling Environment for the U.S. EPA HELP Model for Evaluating and Optimizing Landfill Designs)

An advanced hydrological modeling system called Visual HELP for Windows 95/98/2000/NT is available for designing landfills, forecasting leachate mounding, and assessing potential leachate contamination. In order to construct the model and assess the modeling outcomes, Visual HELP combines the most recent iteration of the HELP model with a simple user interface and strong graphical elements. The user-friendly interface of Visual HELP and its adaptable data handling techniques make it simple to access both the fundamental and more complex HELP model capabilities. This fully integrated modeling HELP environment enables the user to perform sophisticated model simulations, view results in full color and high quality, and prepare graphics and report materials. It also automatically generates statistically reliable weather data and lets you develop your own weather data.

22.5.9 Visual MODFLOW (Integrated Modeling Environment for MODFLOW, MODPATH, MT3D)

Employing MODFLOW-2000, MODPATH, MT3DMS, and RT3D, Visual MODFLOW offers expert 3D groundwater flow and pollutant transport modeling. The most comprehensive and potent graphical modeling environment on the market is provided by Visual MODFLOW Pro, which seamlessly combines the standard Visual MODFLOW program with WinPEST and the Visual MODFLOW 3D-Explorer. This groundwater modeling environment's full integration enables:

- Create the model grid, attributes, and boundary conditions graphically.
- Create a two- or three-dimensional visualization of the model input parameters.
- Run simulations of the transmission of contaminants, path lines, and groundwater movement.
- Calibrate the model automatically with WinPEST or manual techniques.
- Utilize the Visual MODFLOW 3D-Explorer to visualize and analyze the modeling results in three dimensions.

Given the high unpredictability and swift growth of groundwater models, a new more advanced model might frequently take the place of a model that was previously used. A fresh distribution of the parameters may also be necessary for the conceptual model revision and mesh regeneration. As a result, it's critical that model data (information) be kept separate from a particular model, preferably in GIS-based databases. The setup and modification of models are now simpler and more time-efficient thanks to significant advancements in user-friendly GIS and database servers. FEFLOW is an example of such a model since it combines mathematical modeling with GIS-based data exchange interfaces. Natural and artificial stress, as well as the parameters, dimensions, and physico-chemical characteristics of every aquifer taken into account by the model, are all input data for groundwater models. The amount of data required increases dramatically when the numerical approximation (solution) is refined. Common variables like transmissivities, aquitard resistances, abstraction rates, groundwater recharges, surface water levels, etc. are used as input data for aquifers. Groundwater levels, fluxes, velocities, and changes in these parameters as a result of stress applied to the model are the most often seen output data.

22.6 APPLICATION OF ARTIFICIAL INTELLIGENCE IN GROUNDWATER MODELING

Groundwater level (GWL) estimation and analysis in aquifers is a crucial and beneficial task in managing groundwater resources, and knowledge of the GWL types can be used to gauge groundwater accessibility. The GWL variants in wells provide a quick snapshot of the impact of groundwater improvement, and valuable information about aquifer components is frequently included in the consistently recorded GWL time series (Butler *et al.*, 2013). Therefore, it is crucial for water managers and specialists to show and anticipate GWL in order to evaluate and assess groundwater resources and maintain balance between supply and demand.

The primary tool for GWL showing is typically a reasonable or actual based model; nevertheless, these models have a few functional limitations, such as the need for a large amount of information and information boundaries. In general, knowledge is constrained, and obtaining precise expectations is more important than understanding fundamental systems; as a result, black-box artificial intelligence (AI) models can be a viable alternative.

Although there are many methods for displaying and anticipating GWL in aquifers, including theoretical, physical, mathematical, measurable, and others, recently, AI methods have been used for their simplicity and satisfactory results, and many researchers have looked at the display of AI models for GWL displaying in various contexts.

22.6.1 Artificial Intelligence Methods for GWL Modeling

i. **Artificial neural networks (ANN) for GWL modeling**: ANNs are computational models that have biological neural networks as their life force. They can be used to estimate functions that are frequently obscure or to project future benefits of potentially complex time series based on historical data. ANNs are constructed from simple parts that operate in unison. Similar to nature, the relationships between components typically determine an organization's capacity. A typical ANN consists of a number of neurons (handling components), as well as association routes that link them. The neurons with similar characteristics are grouped together in a single layer. An ANN typically has three distinct layers, namely input, hidden, and output layers. Due to GWL gauging, the input elements used by the information layer are often time series for precipitation, temperature, GWL, and other variables. Each neuron in the hidden and output layers transmits its biased, weighted input through an ideal exchange (enactment) capability to produce an effect. A specific piece of information triggers a distinct objective result because ANNs are programmed with example data. To advance the execution of the organization, preparation entails fine-tuning the flexible organizational boundaries (referred to as loads and predispositions). With various preparation (learning) computations, the preparation cycle should be feasible. The Levenberg-Marquardt (LM) algorithm, the back-propagation (BP) algorithm, the Bayesian regularization (BR) algorithm, the gradient descent with force and versatile learning rate back-engendering (GDX) algorithm, and others are examples of the writing algorithms that require the most time and effort to prepare.

Recent GWL modeling studies have revealed that ANNs may be a promising alternative for reasonable approaches. In one of the earliest research, Coulibaly *et al.* (2001) considered three different ANN model types using GWL, precipitation, and temperature time series as the contributions of models to recreate typical month-to-month GWL in the Gondo spring, Burkina Faso. The output of the simulation showed that the RNN is often superior to the static construction input defers ANN and RBF-ANN. ANN was evaluated for measuring the month-to-month GWL in an unconfined chalky aquifer in Northern France (Lallahem *et al.*, 2005). The main objective was to reproduce the GWL in a selected piezometer using the data from 13 piezometers' GWLs, precipitation, mean temperature, and precipitation. The simulations showed how useful using MLP models are. On the Greek island of Crete, Daliakopoulos *et al.* (2005) tested seven different ANN models with various designs and preparing algorithms for month-to-month GWL anticipating. The historical GWL, temperature, precipitation, and stream flow were the input variables.

ii. **Adaptive neuro-fuzzy inference system (ANFIS) for GWL modeling**: The flexible neuro-fuzzy inference system combines an adaptable neural network (AN) and a fuzzy inference system (FIS), and therefore, it may be able to combine the benefits of two different approaches into a single framework. In order to generate a number of fuzzy rules with adequate participation capabilities (MFs) from the preset information yield matches, Jang (1993) developed engineering and a learning technique for the ANFIS that makes use of a neural network learning algorithm. The FIS is compared to a collection of fuzzy when it is determined that it has the ability to learn how to make nonlinear assumptions. For FIS, specifically, there are two methodologies: Mamdani and Sugeno. The differences between these two techniques can be seen in the final section. While Sugeno's methodology uses straight or consistent MFs, Mamdani's methodology makes use of fuzzy MFs. Without attempting to understand the concept of the peculiarities, the ANFIS is an AI technique with adaptable numerical development that is suitable for identifying complicated nonlinearity and vulnerabilities caused by irregularity and imprecision between variables. This method is suitable for accurately approximating any truly persistent capability on a reduced set. In this way, a useful planning that roughly approximates the course of assessment of the inside framework boundary may be built by ANFIS in boundary assessment/determining where the given information are to such an extent that the framework partners quantifiable framework factors with an inside framework boundary. Supplementary data on ANFIS can be found in Jang (1993).

Kholghi and Hosseini (2009) used common kriging and ANFIS for the spatial addition of GWL in an unconfined aquifer in Qazvin, Iran, in the space of GWL demonstrated using ANFIS. For building and testing the models, they used the GWL data from 95 wells. The ANFIS models made use of the Gaussian MF. The findings demonstrated that a shape plot of isopieze lines evaluated by ANFIS was more effective than those evaluated using kriging.

Jalalkamali *et al.* (2011) analyzed how effectively ANFIS and ANN could predict the GWL of a different well in Iran's Kerman plain using various combinations of monthly temperature, precipitation, and GWLs from two adjacent wells. Results indicated that the optimal information mix for predicting GWL was the present and 1-month-ahead GWLs of the well and its neighboring wells, and that the ANFIS models using Gaussian MF performed better than the ANNs.

iii. **Genetic programming (GP) for GWL modeling**: The GP is a hypothesis for a genetic algorithm (GA), and it is an algorithm for natural advancement driven by Darwinian hypotheses about random choice and natural selection. The general practitioner (GP) takes into account an underlying population of conditions that are formed haphazardly and are the result of random causes, numbers, and works. The function provides arithmetic operators (+, −, ×, ÷) as well as additional numerical functions (e.g., sin, cos, and so forth) or client-defined expressions, which should be chosen in light of some cycle understanding. The underlying demographic is then used in a developmental cycle to characterize a fitness function and evaluate the fitness of the created projects. The root mean square error (RMSE) between anticipated and observed data is frequently used as the fitness function in problem prediction. Through the use of two genetic administrators, crossover and mutation, the projects that best match the data are then selected to produce better programs. The evolution process is continually repeated and directed toward discovering articulations that convey the data and provide the best demonstration of the model.

The capabilities of GP, ANFIS, ANN, support vector machine (SVM), and ARIMA approaches for routine GWL determination in Korea were studied by Shiri *et al.* (2013). As the models' contributions, the GWL, precipitation, and evapotranspiration data were used. The fitness function for GP models was the root relative squared error. The results demonstrated that GP models were more analytically sound than other models.

In three wells in Iran's Karaj plain, Fallah-Mehdipour *et al.* (2013) examined how well the GP and ANFIS could predict and recreate month-to-month GWLs. The contributions of the models were used to calculate the precipitation, dissipation, and GWLs. They have noted that the GP's fitness function was perceived as an error criterion, but they haven't specified what kind of error it is. Results showed that when compared to the ANFIS models, the GP models made fewer common errors.

iv. **SVM for GWL modeling**: A quantifiable AI theory is the SVM. Although the information vectors supporting the model's structure are picked throughout the model preparation process, its structure is not predetermined (Vapnik, 1998). This AI method depends on increasing the likelihood of identifying a hyper-plane that distinguishes between two classes in characterization. In an infinite layered space, an SVM creates hyper-planes that can be used for layout, relapse, or other purposes. By describing dot items in terms of a portion capability selected to suit the problem, the mappings used by SVM plans are meant to ensure that dot items may be registered successfully about the components in the first space. The SVM can also be used as a relapse prevention method. With a few minor differences, the help vector relapse (SVR) technique uses the same criteria for categorizing as the SVM. A proper setting of some boundaries and the kernel function are prerequisites for the SVR speculation execution. The SVR bounds regulate the complexity of the expectation (relapse) model and handle a few constants like regularization stable and kernel function consistent. To more accurately perform the relapse task, the kernel function modifies the dimension of the information space. Vapnik presents a comprehensive numerical sketch of SVM. It was initially developed for characterization but was later expanded to address forecasting problems and, to this extent, was used in hydrology-related projects.

Yoon *et al.* (2011) developed ANN and SVM models to account for a 6-hourly time step while predicting GWL vacillations of two wells at a beachfront aquifer in South Korea. The contributions of the models were chosen to be the previous GWL, precipitation, and tide level. The historical GWL was considered to be the best informational variable for the review site, and tidal level was considered to be more significant than precipitation. The results demonstrated that the SVM was superior to the ANN in this instance. Yoon *et al.* (2016) dealt with work on the execution of ANN and SVM models for the expectation of daily GWL due to precipitation using a weighted error function approach. GWL and data on South Korean precipitation were the informational factors. The models' correlation demonstrated that the recursive forecast display of the SVM was dominant to ANN.

v. **Hybrid AI techniques for GWL modeling**: In order to increase the capabilities of the AI strategies, some hybrid modeling approaches that incorporate specific information pre-processing as well as join different AI processes have also been developed in recent years. This is because it has been discovered that the AI models have some limitations with the non-linear and non-fixed processes. The development of these models is made more interesting by combining multiple AI techniques in different phases of the process and by using effective techniques for input information pre-handling. The GP technique, for instance, can be used to enhance AI input elements or possibly AI guideline bounds. Another model combines AI techniques for spatiotemporal GWL showing with geostatistical techniques, such as kriging. In articles, hybrid AI-geostatistical models have been applied to take advantage of the true potential of geostatistical tools for spatiotemporal GWL simulation.

REFERENCES

Butler, J.J., Stotler, R.L., Whittemore, D.O., Reboulet, E.C. 2013. Interpretation of water level changes in the high plains aquifer in western Kansas, *Groundwater*, vol. 51, no. 2, pp. 180–190.

Coulibaly, P., Anctil, F., Aravena, R., Bobee, B. 2001. Artificial neural network modeling of water table depth fluctuations, *Water Resour. Res.*, vol. 37, no. 4, pp. 885–896.

Daliakopoulos, I.N., Coulibaly, P., Tsanis, I.K. 2005. Groundwater level forecasting using artificial neural networks, *J. Hydrol.*, vol. 309, pp. 229–240.

Fallah-Mehdipour, E., Bozorg Haddad, O., Marino, M.A. 2013. Prediction and simulation of monthly groundwater levels by genetic programming, *J. Hydro-Environ. Res.*, vol. 7, no. 4, pp. 1–8.

Anonymous. 2020. Groundwater Modeling. https://www.waterloohydrogeologic.com/learning-groundwater-modeling/

Jalalkamali, A., Sedghi, H., Manshouri, M. 2011. Monthly groundwater level prediction using ANN and neuro-fuzzy models: a case study on Kerman plain, *Iran. J. Hydroinform.*, vol. 13, no. 4, pp. 867–876.

Jang, J.S.R. 1993. Adaptive-network-based fuzzy inference system, *IEEE Man Cybernetics.*, vol. 23, no. 3, pp. 665–685.

Kholghi, M., Hosseini, S.M. 2009. Comparison of groundwater level estimation using neuro-fuzzy and ordinary kriging, *Environ. Model. Assess.*, vol. 14, no. 6, pp. 729–737.

Kumar, C.P. 2001. Introduction to Groundwater Modeling. http://nihroorkee.gov.in/sites/default/files/uploadfiles/CPK_GW_Modelling.pdf

Lallahem, S., Mania, J., Hani, A., Najjar, Y. 2005. On the use of neural networks to evaluate groundwater levels in fractured media, *J. Hydrol.*, vol. 307, pp. 92–111.

Shiri, J., Kisi, O., Yoon, H., Lee, K.K., Nazemi, A.H. 2013. Predicting groundwater level fluctuations with meteorological effect implications – a comparative study among soft computing techniques, *Comput. Geosci.*, vol. 56, pp. 32–44.

Vapnik, V.N. 1998. *Statistical Learning Theory*, Wiley, New York, 736.

Yoon, H., Hyun, Y., Ha, K., Lee, K.K., Kim, G.B. 2016. A method to improve the stability and accuracy of ANN- and SVM-based time series models for long-term groundwater level predictions, *Comput. Geosci.*, vol. 90, pp. 144–155.

Yoon, H., Jun, S.C., Hyun, Y., Bae, G.O., Lee, K.K. 2011. A comparative study of artificial neural networks and support vector machines for predicting groundwater levels in a coastal aquifer, *J. Hydrol.*, vol. 396, no. 1, pp. 128–138.

McDonald, M.G., and Harbaugh, A.W., 1988, A modular three-dimensional finite-difference ground-water flow model: U.S. Geological Survey Techniques of Water-Resources Investigations, book 6, chap. A1.

Appendix
Measurement Units and Their Conversion

Length

1 inch	=	2.54 cm
1 mile	=	1,609.344 m
1 foot	=	12 in
1 yard	=	3 ft
1 international nautical mile	=	1,852 m
1 link	=	0.66 ft
1 chain (Gunter's Chain)	=	66 ft
		20.1168 m
		100 links
1 rod, pole or perch	=	16.5 ft
40 rods	=	1 fur
		660 ft
1 furlong	=	201.168 m
1 rope	=	20 ft
1 fathom	=	6 ft

Area

1 square foot	=	144 sq in
1 square yard	=	9 ft^2
1 hectare	=	10,000 m^2
1 cubic meter	=	1,000 L
1 liter	=	100 cm^3
1 barrel (oil)	=	159 L
1 gallon (USA)	=	3.78 L
1 cusec	=	1 ft^3/sec
1 cumec	=	1 m^3/sec

Weight

1 slug	=	14.594 kg
1 pound	=	0.4536 kg
1 ton (metric)	=	1,000 kg
1 quintal (metric)	=	100 kg

Velocity

1 knot	=	0.514 m/sec
1 mach (SI standard)	=	295.0464 m/sec

Pressure

1 atmosphere	=	101,325 Pa
		101,325 N/m^2
		1.101329 bar
1 pascal	=	1 N/m^2
1 bar	=	100,000 Pa

Energy

1 calorie	=	4.18 J
1 joule	=	1 N-m
1 BTU	=	1,055.056 J

Power

1 watt	=	1 J/sec
1 horsepower	=	745.7 watt
1 ton (refrigeration)	=	3.52 KW

Force

1 newton	=	100,000 dyne
1 kilogram-force	=	9.8 N
1 pound-force	=	4.448 N
1 poundal	=	0.138254954 N

Angle

1 radian	=	180°
1 degree	=	60 minutes
1 revolution	=	360°
1 right angle	=	90°

Temperature

1 degree Celsius	=	274.15 K
T (Kelvin)	=	T(°C)+273.15
T (Rankine)	=	T(°F)+459.67

Data Storage

1 byte	=	8 bit
1 word	=	2 bytes
I megabyte	=	1,024 KB

Glossary

Alluvial: Geological word used to describe an origin connected to surface water flowing on land. Consider alluvial sediments or deposits, which are those that are located along current or former stream systems and are typically made up of gravel, sand, silt, and/or clay.

Aquifer: An area of groundwater-containing porous or fractured rocks or highly permeable unconsolidated porous sediments that is hydraulically continuous. It has the potential to produce usable amounts of groundwater.

Aquitard: A body of poorly permeable rocks filled with groundwater, through which large amounts of groundwater may still flow, albeit slowly.

Aquiclude: Poorly permeable formations that are filled with groundwater but through which nearly little groundwater flows.

Aquifuge: An impermeable structure that does neither convey nor contain water.

Arid region: Low precipitation areas known as "arid regions" are characterized by a significant lack of water resources, which can impede or even prevent plant growth. This phrase is used in agriculture to describe excessively arid regions where crops cannot be produced without irrigation.

Artesian: If a well is drilled into the area where such pressure is prevalent (an "artesian aquifer"), groundwater will be able to rise above the local ground surface. In other words, the artesian aquifer's piezometric groundwater level is placed above ground. A "flowing well" or "artesian well" is a well that draws water from an artesian aquifer.

Base flow: The continuous stream flow component that lasts for a considerable amount of time following the last rain. It is occasionally referred to as "dry-weather flow" and is mostly fueled by groundwater system discharge.

Basement: A complex of metamorphic and igneous rocks, often Precambrian or Palaeozoic in age, that lies beneath all sedimentary formations. Basement rocks are the oldest rocks known to exist in a particular place.

Bedrock: Solid rock that has not weakened or been significantly altered is called bedrock (thus it is rather dense, and usually of low porosity).

Black water: Wastewater that still has sanitary traces in it.

Blue water: Groundwater and natural surface water.

Brackish water: Water with dissolved solids in it that ranges in concentration from 1,000 to 10,000 mg/L is referred to be brackish water.

Borehole: A hole drilled into the ground or the upper crust of the earth. Its goal is typically either geological exploration or the construction of a well for extracting or injecting liquids, and its diameter is typically tiny (e.g. water).

Cainozoic: A geological period that lasted roughly from 66 million years ago to the present day. It is frequently used as an adjective to denote a geological formation's age.

Capillary fringe: The area immediately above the water table known as the capillary fringe is where the interstices are totally filled with water but are still subject to air pressure (due to suction forces).

Carbon cycle: The exchange of carbon between the biosphere, pedosphere, geosphere, hydrosphere, and atmosphere of the Earth occurs through the biogeochemical cycle known as the carbon cycle.

Carbonate rock: A rock made up of carbonate minerals, particularly dolomite and limestone.

Catchment area: The entire area with a single outlet for the discharge of its surface water. It is also known as a river basin or a watershed.

Climate: A statistical synthesis of the local weather over a lengthy period of time, including mean values, variances, probabilities of extreme values, etc (usually taken as 30 years).

Climate change: Long-term modification of the climate (represented by a shift in the averages and/or variability of climatic variables' long-term statistical features).

Collector well: Well equipped with horizontal tube drains that are positioned in various radial orientations to improve the collector well's effective radius.

Confined aquifer: This is an aquifer that is directly covered by an impermeable or nearly impermeable formation and is totally saturated (i.e., pressure everywhere is greater than atmospheric pressure) (confining bed). The aquifer cannot interact with the atmosphere or surface water bodies directly because of the restricting bed (except for surface water bodies that intersect the aquifer).

Confining bed: A fully saturated aquifer is described as being covered by a confined bed, which is an impermeable or poorly permeable formation.

Connate water: It is water that was trapped in a sedimentary rock's pores when it was forming.

Consolidated:

1.: Geological word that denotes that, in theory, a rock forms as a solid mass rather than an accumulation of loose, uncemented substances like gravel, sand, silt, and clay (the latter are unconsolidated sediments).

2.: A word from soil mechanics that denotes that, in reaction to an external force or a drop in hydrostatic pressure, the volume of an earth layer has dropped (mostly because of lower porosity).

Consumptive usage: Refers to the portion of water consumed that is not redirected to aquifers, streams, or oceans but instead is either absorbed into goods or living things or released as vapor into the atmosphere.

Cretaceous: It is the Mesozoic Era's period in geological history. Frequently used as an adjective to denote a geological formation's age.

The Darcy rule: The volume of flow traveling through a section in a porous medium at a particular time interval is proportional to the hydraulic gradient and the hydraulic conductivity of the saturated medium, according to one of the key physical laws in groundwater hydraulics.

Delta: When a river empties into an ocean, sea, estuary, lake, reservoir, another river, or a flat, arid terrain, it forms a landform known as a delta, which is characterized by the deposition of a significant portion of the sediments carried by that river (alluvial sediments).

Depletion: A reduction in the amount of groundwater that is stored in an aquifer (also: reduction of the stored water volume of any other component of the local hydrological cycle).

Deposit: Sediment is geologically referred to as deposit.

Domestic water use: The use of water for drinking and other home needs, in workplaces, and for public water services that are not related to agriculture or industry. The consumers themselves may withdraw the water or it may be provided via municipal or public water delivery systems (self-supply).

Drainage: The natural or induced evacuation of surplus water.

Drawdown: A reduction in groundwater level or piezometric surface brought on by groundwater abstraction (including not only pumping, but also outflow from an artesian well or discharge from a spring).

Ecosystem: An ecosystem is a dynamic system made up of populations of plants, animals, and microorganisms as well as their non-living surroundings.

Endorheic: Adjective suggesting that water from the location, aquifer, or stream in question is released directly into the atmosphere rather than the open sea and seas. Endorheic aquifers are either evaporated in confined depressions or drained by streams in endorheic river basins. Basins of endorheic rivers or lakes are frequently referred to as "closed basins."

Evaporite: It is a chemical sedimentary rock made up of minerals that evaporated water precipitated.

Evaporation:

1. The physical process through which water vapor is released into the atmosphere.
2. The volume of liquid that evaporated. Potential evaporation is the term for the greatest rate of evaporation from a water surface.

Evapotranspiration:

1. The release of water into the atmosphere through evaporation from the ground, from water surfaces, and also from plant leaf surfaces (transpiration).
2. The volume of water evaporated through transpiration. (Potential evapotranspiration is the highest rate, which is the rate of evapotranspiration from a completely established vegetation cover, well supplied with water.)

Exhaustion: Depletion that has progressed to the point where no more resources are exploitable.

Facies: A rock's collection of properties that reveal the specific conditions under which it developed and set it apart from other facies in the same rock formation. For sedimentary rocks, it describes the specific deposition environment, while for metamorphic rocks, it describes the precise range of pressure and temperature where metamorphism took place.

Fissures: Secondary interstices that developed after the formation of the rock.

Flux: A numerical measurement of the flow rate across a certain surface.

Flux density: Flux per square meter.

Folded mountains: Mountains that are formed by the bending of an initially flat earth structure, typically a series of sedimentary strata. The crust's vertical or horizontal forces may have caused the deformation.

Formation: A group of rocks or unconsolidated sediments that are or were once horizontally continuous and share some recognizable lithological characteristics

Fossil groundwater: Groundwater that has been stored inside a rock formation over a long period of time (under climatic and/or geological conditions distinct from those of today) and that cannot be replenished in the present.

Fracture: When forces in the crust exceed a particular threshold, cracks in rock masses result (below this critical point folding may occur). Joint fractures and fault fractures are the two types of fractures. A crack along which there hasn't been any displacement is called a joint. A fracture is referred to as a fault if there is relative displacement of rocks on both sides of the fracture and parallel to it.

Free aquifer: Phreatic aquifer, unconstrained aquifer.

Freshwater: Freshwater is water with a low concentration of dissolved solids (usually less than 1,000 mg/L).

Geyser: A geyser is a hot spring that compulsorily shoots steam and hot water into the air. The heat is considered to be caused by groundwater coming into touch with magma masses.

Geothermal: Concerning the earth's crustal heat energy.

Graben: Geological word for a piece of the earth's crust that has been displaced and is bounded by parallel faults, or "normal faults."

Green water: Green water is precipitation that either stores itself in the soil or briefly lingers on top of the soil or vegetation rather than evaporating or replenishing groundwater. This portion of the precipitation eventually evaporates or seeps through plants.

Gray water: Gray water is contaminated water that is produced when water is used for purposes other than hygienic ones.

Groundwater: It is subsurface water that has a pressure that is greater than or equal to that of the surrounding atmosphere. In other terms, it is defined as phreatic level or water below the water table.

Groundwater development: It is sometimes referred to as groundwater exploitation and is defined as human activity that involves removing groundwater and making it accessible for useful purposes.

Groundwater discharge: Water that has left a groundwater system is called a groundwater discharge.

Groundwater level: The height to which the groundwater in a piezometer attached to a point in the groundwater domain will or will not rise. It is a variable that is time-dependent, varies from one point to another within the groundwater domain, and represents the potential energy of the groundwater at any given place (in meters of water column relative to a selected topographic reference level).

Groundwater mining: Taking up groundwater that won't be replenished is called as groundwater mining. It is also referred to as fossil or non-renewable groundwater.

Groundwater recharging, renewing, and replenishing:

1.: The process of water entering an aquifer or groundwater system.
2.: The process of this inflow's flux.

Groundwater runoff: Groundwater from springs or diffuse seepage into the stream bed contributes to a portion of the stream flow.

Groundwater table: Surface in an aquifer determined by the phreatic levels (i.e., surface of atmospheric pressure within an unconfined aquifer).

Humid region: A region is considered humid if there is an excess of precipitation compared to potential evapotranspiration. In other words, the area's water supply is adequate to support plant growth and development without irrigation.

Hydraulics: Hydraulics is the branch of practical science that studies how water moves through pipes, open waterways, and porous media. The adjective is hydraulic.

Hydraulic conductivity: The ability of a porous medium to convey water is known as hydraulic conductivity. Its equivalent is permeability.

Hydraulic gradient: Water level change (piezometric level) per unit of horizontal distance is referred to as a hydraulic gradient.

Hydrograph: A hydrograph is a graph that shows the flow of water through time.

Hydrological cycle: The circulation of water at or near the Earth's surface is known as the hydrological cycle (in which water is moving between the subsystems atmosphere, surface water, unsaturated zone, groundwater, and oceans).

Hydrological over-abstraction: It is the term for extraction from a renewable groundwater resource that is excessive in comparison to the underlying aquifer's hydrological budget. A new dynamic equilibrium cannot be developed in the intermediate or long run due to the high abstraction intensity.

Hydrology: The study of the water cycle on land is known as hydrology.

Hydrosphere: Waters on the Earth are referred to as the hydrosphere, as opposed to rocks (lithosphere), living organisms (biosphere), and air (atmosphere).

Igneous rock: A rock created when magma solidifies.

Impermeable, impervious: Water cannot pass through impermeable or impervious materials (property of a porous medium).

Infiltration:

1. The act of allowing water to enter a fissured or porous formation.
2. The amount or pace of water infiltration.

Infiltration gallery: A man-made structure that draws water from the earth and transports it to the surface without the use of external energy (the system is based on flow by gravity).

Intensive development of groundwater: Groundwater development that is so extensive that it considerably alters the aquifer's or system's natural flow is referred to as intensive groundwater development (significant change of flow regime).

Interface: The point of contact between two different bodies, such as between fresh and salt groundwater. The transition between salty and fresh groundwater may be relatively abrupt or comprise a thick transition zone, depending on the local conditions (zone where fresh- and groundwater are mixed).

Interstices (voids): Unconsolidated subsurface sediments and rock formations containing open spaces (i.e., the area not occupied by solid matter) that allow fluids (such as air, water, oil, and gas) to pass through or be stored below earth. Primary interstices (mostly pores) and secondary interstices, which developed after the rock was formed, are two categories of interstices (mainly fissures, sometimes widened by dissolution). Fissured or porous are the related rock/sediment classifications, respectively. Porosity is a term used to describe the proportion of a rock's bulk volume that is filled by interstices (pores and fissures combined).

Irrigation: Irrigation is the artificial application of groundwater or surface water to soil or land, typically with the goal of fostering crop development.

Karst: A phenomenon that results from the disintegration of solid rock, primarily limestone, that generates irregularities in the ground surface and drainage features (both internal and exterior).

Land subsidence: It is the sinking elevation of the ground surface that can happen when water is drained or extracted from subsurface strata, such as an aquifer. This can cause sediment to compress as a result of the lower water pressure.

Lithology: It is the area of earth science that deals with categorizing and describing various forms of rock and their physical characteristics.

Marine: : A geological phrase denoting a sea or ocean-related origin.

Marine regression: A seaward displacement of the land-ocean divide that lasts for a protracted period of geological time.

Marine transgression: A long-lasting, geologically significant shift of the landward boundary between the sea and the land.

Mesozoic: A geological period that lasted from around 250–66 million years ago. Frequently used as an adjective to denote a geological formation's age.

Metamorphic rock: Rocks that have undergone metamorphism have altered mineralogy, texture, or composition as a result of pressure, temperature, or the addition or subtraction of chemical elements.

Mineral water: It is natural water that has more dissolved salts than a specified threshold and is frequently thought to have medicinal benefits.

Mining: The process of using up a non-renewable natural resource's supply.

Monitoring: It is the process of repeatedly and methodically observing a specific characteristic or phenomena.

Natural resource: A typical definition of a natural resource is "anything found in nature that is necessary or helpful to humans." This usually refers to a substance or "raw material." Common examples include sun energy (a permanent resource), water and forests (renewable resources), or minerals and fossil fuels (non-renewable resources).

Non-renewable groundwater: A body or unit of groundwater that receives little to no recharge under the current climatic circumstances.

Overexploitation: Excessive levels of intensive exploitation. There is no universally accepted definition of groundwater overexploitation among experts in the field, and it is frequently unclear exactly by what standard the exploitation is deemed excessive. Overexploitation in this context is understood to be extensive exploitation marked by a less favorable balance between positive side effects and negative side effects than would have been the case at a lower rate of exploitation. (Consider the hydrological over-abstraction.)

Paleozoic: A geological era that spanned around 590–250 million years ago. Frequently used as an adjective to denote a geological formation's age.

Permeable, pervious: Having the ability to transport large amounts of water (property of a porous medium).

Permeability: A porous medium's ability to transfer water. Hydraulic conductivity is a synonym.

Permafrost: A layer of soil or rock where the temperature has consistently been below 0°C for at least a few years.

Phreatic aquifer: It is also known as a "water-table aquifer," is a type of aquifer where the top of the groundwater mass creates a surface that is in direct contact with the atmosphere. This circumstance favors the aquifer's participation in the water cycle. Water-table aquifer, unrestricted aquifer, and free aquifer are synonyms.

Phreatic level: When the water pressure in an aquifer is exactly equal to the atmospheric pressure in the area, it is said to be at a phreatic level. It is the depth that a shallow well in such an aquifer recorded.

Phreatophytes: Water-loving plants known as phreatophytes receive groundwater either continuously or sporadically through the capillary fringe of their roots.

Piedmont: A region created by the byproducts of erosion that lies at the base of a mountain or mountain range.

Piezometer: A piezometer is a tool or well used to gauge the amount of potential energy present at a particular water body point (stream, lake, aquifer, aquitard, etc.).

Piezometric level: The elevation to which water in a piezometer attached to a spot in an aquifer would rise is known as the piezometric level. Synonyms include hydraulic head, piezometric head, and potentiometric level.

Piezometric surface: A surface inside an aquifer that is defined by piezometric levels that correspond to a specific stratum or topographic level. Potentiometric surface is a synonym.

Plateau: Extensive, horizontal, relatively level highland region known as a plateau that is typically higher than the surroundings and is bordered on at least one side by steep slopes. Alternative names: meseta, altiplano (the latter usually restricted to a highland intermontane plateau).

Platform: A sizable, horizontal, relatively level region that is typically higher than its surroundings.

Porosity: The amount of holes or pores compared to the material's overall volume is known as the material's porosity.

Precambrian: A geological epoch that spanned from 4,000 to 590 million years ago. Frequently used as an adjective to denote a geological formation's age.

Precipitation:

1.: Water vapor condensation products that fall from clouds or are deposited from the air on the ground, such as rain, hail, snow, and dew.

2.: The amount of water that falls as rain, hail, snow, etc. on a unit of horizontal surface over the course of one unit of time.

Quaternary: The most recent geological epoch, part of the Cenozoic Era. Frequently used as an adjective to denote a geological formation's age.

Recharging area: An aquifer's primary recharge area is known as recharging area.

Regolith: Any solid substance resting on top of bedrock, including soil, alluvium, and broken pieces of bedrock that have weathered.

Renewable groundwater: : A body of groundwater that replenishes as a result of the geological and climatic circumstances that exist today.

Reserves: Storage volume is a reserve.

Resilience: A system's resilience is its capacity to bounce back from an undesirable situation (in particular, the ability to remain in or return to a state of dynamic equilibrium).

Return flow: Any flow of water that has been extracted and returns to a stream channel or to the groundwater reserves after consumption.

River basin: A river's catchment area.

Runoff: The portion of precipitation that appears as stream flow is known as runoff. Interflow is a delayed flow that passes partially through the top soil layers, groundwater runoff is the outflow of groundwater to streams, and surface runoff is the outcome of overland flow during and after a storm or precipitation event. Often, these three components are differentiated (forming the essential part of baseflow).

Runoff coefficient: The ratio between the amount of runoff and the amount of rainfall during the event that induced runoff is known as the runoff coefficient.

Safe yield: Groundwater flow that can be drained from an aquifer without having an unfavorable impact.

Saline water: Water that has dissolved minerals in amounts greater than 10,000 mg/L is referred to as saline water.

Salinity: Salt concentration in water is known as salinity.

Seawater invasion: It is also known as salt water intrusion and is the invasion of a freshwater body by seawater (often seawater) (either surface water or a groundwater body).

Saprolite: A section of the crystalline basement aquifers' weathered/decomposed zone (regolith), which is rich in clay minerals. It has a much higher storage capacity compared to the lower, more permeable fissured zone while having a poor hydraulic conductivity (saprock).

Saturated zone: It is defined as a region of the subsurface water-bearing deposit where water fills all spaces, big and small.

Scale: The proportion of a map's distances to their actual distances (e.g., 1: 50,000 or 1:1,00,0 000). A small-scale map may show a sizable area for a given map size, but with little or no detail.

Secondary groundwater sources: Additional sources of groundwater recharge, such as irrigation water losses or return flows (byproducts), as well as artificial recharge, are referred to as secondary groundwater resources.

Seepage:

1.: Water moving slowly through a porous media.

2.: Water loss caused by water entering a porous medium (infiltration) or escaping from one along a surface or line (exfiltration).

Semi-arid region: An area with low annual precipitation that is characterized at least seasonally by a lack of water supply, to the point where plant growth and development are hampered, at least during the dry season. This phrase is used in agriculture to characterize dry locations where farming is either impossible or just marginally profitable without irrigation.

Semi-confined aquifer: Confined aquifers that have either the confining bed on top of the aquifer or a poorly permeable basal formation underlying the aquifer have enough permeability (aquitards) to for water exchange between the aquifer and the domains above or below it are referred to as semi-confined aquifers.

Semi-permeable bed: A layer of the earth called a semi-permeable bed has a limited ability to transport groundwater. Substitute is an aquitard.

Sinkhole: A depression in karst terrain that serves as the place where surface water flow disappears. Usually, the dissolution of karstified limestone or dolomite near the ground's surface transforms it into a funnel-shaped cavity (doline).

Soil moisture: Water found in soil pores above the water table is known as soil moisture.

Speleology: It is the study of caves.

Spring: The season when there is a lot of natural groundwater evaporation.

Storage, stored volume, and stock: These all refer to the amount of water in a given hydrological system component (e.g. aquifer, stream, lake, soil).

Stream: Water body flowing in a surface channel naturally. A general term that refers to natural waterways such rivers, rivulets, brooks, and others.

Subsidence: The height of the earth sinking as a result of either natural or human-caused factors. It happens when sediment is compressed as a result of decreased water pressure, which can happen, among other things, when water is drained or extracted from subterranean strata, such an aquifer or an aquitard.

Surface water: Water that is present on the Earth's surface, such as that found in lakes, rivers, and streams.

Sustainable yield: The amount of groundwater that can be removed from an aquifer without producing unfavorable side effects, particularly without putting the aquifer's hydrological budget in a permanent state of disequilibrium.

Synonyms: Piezometric level, piezometric head, hydraulic head, groundwater hydraulic potential.

Terrestrial: It is associated to land masses as opposed to marine and atmospheric.

Tertiary: A Cainozoic Era-related geological time period. Frequently used as an adjective to denote a geological formation's age.

Texture, rock texture: A rock's physical characteristics that are determined by the size, shape, and arrangement of the minerals that makes up its composition.

Trans-boundary aquifer: Aquifers that cross political boundaries to connect two or more political entities are known as trans-boundary aquifers.

Phreatic aquifer: Unconstrained aquifer.

Unconsolidated sediments: Sediments made up of an uncemented matrix of loose particles.

Unsaturated zone: Area below the ground's surface where air and water are mixed in equal amounts. The water table or the top of a constrained aquifer are both above this zone. In the unsaturated zone, the water pressure is lower than the atmospheric pressure.

Virtual water: Water used in the creation of goods or services is referred to as virtual water.

Water budget: A plan that details the inflows, outflows, and changes in the volume of water stored for a given water system (a river basin, an aquifer, a soil system, or an area) during a given time period. In theory, the difference between inflows and outflows equals the change in the volume that has been stored.

Water cycle: The circulation of water at or near the Earth's surface is known as the water cycle (in which water moves between the subsystems atmosphere, surface water, unsaturated zone, groundwater, and oceans).

Water footprint: The entire amount of freshwater utilized to generate the commodities and services that a person, a community, or a corporation uses to operate.

Well: An artificial structure, typically vertical and cylindrical in shape, used to access groundwater and lift it to the surface for abstraction.

Wetland: A wetland is a region where the ground is perpetually or seasonally wet (swamp, marsh, peatland, shallow lake).

Withdrawal:

1. The act of temporarily or permanently drawing water from a source.
2. The quantity of water used in this operation (usually expressed as a volume per unit of time).

Index

Printed in the United States
by Baker & Taylor Publisher Services